OCCUPATIONAL
PHYSIOLOGY

T0132568

OCCUPATIONAL PHYSIOLOGY

Edited by
ALLAN TOOMINGAS • SVEND ERIK MATHIASSEN
EWA WIGAEUS TORNQVIST

CRC Press
Taylor & Francis Group
Boca Raton London New York

CRC Press is an imprint of the
Taylor & Francis Group, an **informa** business

UNIVERSITY
OF GÄVLE

Published in
cooperation with
the University of
Gävle, Sweden

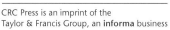

This book was previously published in Sweden as *Arbetslivsfysiologi*, by Studentlitteratur.

CRC Press
Taylor & Francis Group
6000 Broken Sound Parkway NW, Suite 300
Boca Raton, FL 33487-2742

First issued in paperback 2019

© 2012 by Taylor & Francis Group, LLC
CRC Press is an imprint of Taylor & Francis Group, an Informa business

No claim to original U.S. Government works

ISBN-13: 978-1-4398-6696-2 (hbk)
ISBN-13: 978-0-367-38179-0 (pbk)

This book contains information obtained from authentic and highly regarded sources. Reasonable efforts have been made to publish reliable data and information, but the author and publisher cannot assume responsibility for the validity of all materials or the consequences of their use. The authors and publishers have attempted to trace the copyright holders of all material reproduced in this publication and apologize to copyright holders if permission to publish in this form has not been obtained. If any copyright material has not been acknowledged please write and let us know so we may rectify in any future reprint.

Except as permitted under U.S. Copyright Law, no part of this book may be reprinted, reproduced, transmitted, or utilized in any form by any electronic, mechanical, or other means, now known or hereafter invented, including photocopying, microfilming, and recording, or in any information storage or retrieval system, without written permission from the publishers.

For permission to photocopy or use material electronically from this work, please access www.copyright.com (http://www.copyright.com/) or contact the Copyright Clearance Center, Inc. (CCC), 222 Rosewood Drive, Danvers, MA 01923, 978-750-8400. CCC is a not-for-profit organization that provides licenses and registration for a variety of users. For organizations that have been granted a photocopy license by the CCC, a separate system of payment has been arranged.

Trademark Notice: Product or corporate names may be trademarks or registered trademarks, and are used only for identification and explanation without intent to infringe.

Visit the Taylor & Francis Web site at
http://www.taylorandfrancis.com

and the CRC Press Web site at
http://www.crcpress.com

Contents

Preface

A well-functioning society is dependent on a productive working life, which not only provides the working population with an income but also promotes health and well-being. This book, *Occupational Physiology*, is about the working human's physical and mental capacities and needs, and how these correspond to prevailing conditions in current working life. Solid knowledge of these issues is an essential basis for creating a working life that is, at the same time, effective and pursues the vision of "a health-promoting working life for everybody."

This book meets a need for a clear and easy-to-grasp textbook focusing on important issues in occupational physiology. It addresses some of the major public health problems of the present time—musculoskeletal disorders and stress—and describes how these problems are related to conditions in working life. The book explains associations among work, well-being, and health according to recent theories and knowledge within the field. It also presents methods for risk assessment and provides information about relevant health and safety directives. Finally, emphasizing the health-promoting potential of work, advice and guidelines are given on how to arrange for a good working life from the perspective of the working individual, the company, and society as a whole. Thus, the book has its objective in being of practical use.

Occupational Physiology is a textbook for educational programmes devoted to "man at work," for instance, professionals in occupational health services, in occupational and environmental medicine, and in other professions devoted to health promotion in occupational life. It also addresses safety officers, trade union representatives, company management, and other stakeholders having influence on the organization and contents of work.

The book focuses on a number of common, yet stressful, working conditions, such as work requiring considerable physical or mental effort, sedentary or repetitive work, work at inconvenient hours or work in a hot or cold climate. The chapters are written by experts within each area. By referring to situations in working life, the book differs from other textbooks in work physiology, which adhere to a traditional medical–anatomical structure, separating the physiology of the cardio-respiratory system, the nervous system, and so on. The present new approach makes the book easier to grasp, closer to applications in working life, and thus more attractive to the target groups.

The book is intended to be easily accessible to anyone with a secondary education, and does not require any prior specialist knowledge of medicine or physiology. "Fact boxes" in the chapters explain and further deepen the text. References and suggestions for "Further Reading" are at the end of each chapter for the purpose of additional knowledge.

The book is an updated and internationalized translation of a Swedish textbook, *Arbetslivsfysiologi*, published in late 2008 by Studentlitteratur in Lund. The former National Institute for Working Life in Sweden supported the original publication, whereas the translation into English was funded by the University of Gävle.

The original Chapters 2 through 9 were reviewed for their physiological content by Professor Gisela Sjøgaard, currently associated with the Institute of Sports Science and Clinical Biomechanics at the University of Southern Denmark, and we wish to express our sincere appreciation for her contribution. Margareta Bratt Carlström, ergonomist at Avonova Occupational Health Services in Stockholm, has provided us with excellent comments on the different chapters.

Allan Toomingas
Svend Erik Mathiassen
Ewa Wigaeus Tornqvist

Editors

Allan Toomingas, PhD, MD, is a registered psychologist, an associate professor in occupational and environmental medicine, and a senior researcher at Karolinska Institutet, Institute of Environmental Medicine and the Centre for Musculoskeletal Research, University of Gävle, Sweden. He is also a physician at the clinic for occupational and environmental medicine at the Karolinska University Hospital in Stockholm. His major research areas include work-related musculoskeletal disorders, healthy ICT work, and methods of occupational health services. He teaches and organizes educational programmes at the Karolinska Institutet mainly for medical students and specialists in occupational health services.

Svend Erik Mathiassen, PhD, is a professor and research director at the Centre for Musculoskeletal Research, University of Gävle, Sweden. His main research interest lies in physical variation in working life: how to measure "variation," effects on performance, fatigue, and disorders of different types of variation, and interventions in working life promoting or counteracting variation. His interest in exposure variability has also led to extensive research on cost-efficient strategies for collecting and analysing data on physical load. Thus, he has been involved in studies of variation among, for instance, hairdressers, industrial assembly workers, cleaners, flight baggage handlers, office workers, and house painters. He is also currently engaged in a scientific advisory committee formed by the Swedish government to aid in matters related to its work environment policy.

Ewa Wigaeus Tornqvist, PhD, was a professor in ergonomics at KTH, Royal Institute of Technology, School of Technology and Health, and the University of Gävle, Centre for Musculoskeletal Research at the time this book was written. Her major research area is work-related musculoskeletal health. She teaches work physiology, ergonomics, and epidemiology for engineering and technology students and for specialists in occupational health services. From March 2011 she has been a professor in occupational health and dean of the School of Health Sciences, Jönköping University, Sweden.

Authors

Torbjörn Åkerstedt, PhD, is a professor of behavioural physiology and director of the Stress Research Institute, Stockholm University. He is also affiliated with the Department of Clinical Neuroscience at the Karolinska Institutet. His major research focus is on sleep, alertness, stress, and work hours. He has published more than 230 papers in peer-reviewed scientific journals.

Margareta Bratt Carlström, BSc in physical therapy, is a physiotherapist/ergonomist. She was a former senior administrative officer at the Swedish Work Environment Authority, Department of Ergonomics and Workplace Design. She is now a consultant in a private company, Avonova Occupational Health Service Ltd. (Stockholm), where she specializes in tasks regarding the legal framework and provisions for work environment conditions.

Désirée Gavhed, PhD, is a researcher at the Karolinska Institutet, Stockholm, Sweden. She has performed laboratory and field studies in thermal and work physiology for many years at the National Institute for Working Life, Stockholm, Sweden. She teaches thermal and work physiology in occupational safety and health education programs, worked as consultant for the Swedish Work Environment Authority, and was involved in international standardization work in the area.

Karin Harms-Ringdahl, PhD, registered physiotherapist, is a professor in physiotherapy at Karolinska Institutet, Stockholm. She has a combined position at the clinic of physical therapy at the Karolinska University Hospital in Stockholm. Her major research areas are work-related musculoskeletal disorders and movement science. She conducts research and teaches and arranges educational courses at the Karolinska Institutet for physiotherapy students.

Fredrik Hellström, PhD, is a researcher at the Centre for Musculoskeletal Research and the Department of Occupational and Public Health Sciences, University of Gävle, Sweden. His major research areas are physiological responses to repetitive and static work as well as basic mechanisms behind work-related muscle pain. He teaches muscle physiology, motor control, and ergonomics to physiotherapists and students within health education.

Katarina Kjellberg, PhD, is a researcher and lecturer at Karolinska Institutet, Department of Public Health Sciences. Her main research interests concern work-related musculoskeletal health, work ability, sickness absence, ergonomics in the health care sector and cash-register work, and methods and practices for the occupational health services, such as methods for assessment of work technique, work ability, and early occupational rehabilitation. She teaches and arranges educational programs at the Karolinska Institutet for specialists in occupational health services.

Bo Melin, PhD, is a professor in work psychology and the head of the psychology division at Karolinska Institutet. His field of research is within psychobiological stress reactions and health. More recently, he is involved in studies regarding cognitive and emotional capacities in relation to health and achievement in life within a work-related frame. This relatively new field is often addressed as cognitive epidemiology. He is a member of several scientific committees and teaches at a new five-year psychology program at the Karolinska Institutet.

1 Work, Working Life, Occupational Physiology

Allan Toomingas, Svend Erik Mathiassen, and Ewa Wigaeus Tornqvist

Photo: Illar Toomingas

CONTENTS

1

1.1   WORK: A MAJOR PART OF LIFE

Working has always been an integral part of human life. Work has been a precondition for the survival of the individual, the success of companies and business, and the existence of society. In their early environments as hunter–gatherers, people contributed, according to their ability, to the tasks that arose. Most of them probably engaged in a broad range of tasks, but presumably with a marked difference between the genders. There were no specific workplaces, but the tasks were carried out within families as a part of daily life. Gradually, a differentiation arose between "work" and "leisure," which has become permanent in most cultures right up to the present day.

Different individuals and groups eventually developed different competences and specializations—professions were created which varied and vary with regard to both physical and mental demands. The most important "work tool" has, in most cases, been the human body itself. Heavy, yet varying, manual work which dominated in earlier days has presumably had a training effect on fitness and strength as well as endurance. Tools and machines have been developed to increase efficiency and to carry out more sophisticated work that would not have been possible without these aids, and thus the burdens—at least the physical ones—have gradually decreased in most occupations. This industrial development also meant that work in most cases had to be located at specific workplaces where equipment was available.

In today's post-industrial society, one can discern a blurring of the differences between work and leisure, particularly in the knowledge production sector of working life. Many tasks can, with the help of electronic information and communication technology (ICT), be carried out almost anywhere and at any time. Work and leisure therefore merge together once again without boundaries in either time or space, as before industrialization. What cannot be discerned, however, is any retreat from professional specialization to a situation where most individuals help carry out most working tasks. On the contrary, modern working life demands increased specialization, even within different professions. Companies and organizations focus their activities on "core areas" in which they develop their specialized competitiveness. Working life in post-industrial society is therefore characterized in many professions by a protracted strain, if at a low intensity, on the body that does not provide any obvious variation. At the same time, the mental strain is often high. The computer is nowadays the most common piece of equipment used. At the same time, many professions with heavy, uncomfortable, or monotonous loads—in health care and construction, for example—still exist.

Work fills a large part of people's lives. In our society, work, and travel to and from work, takes up more than half of our waking lives during the working week. How work is designed and carried out has considerable significance both for individual performance, health and well-being, as well as for the success of companies and business and the existence of communities.

1.2 WORK, EXPOSURE, AND PHYSIOLOGICAL RESPONSES

Every time someone is faced with a task, they decide—consciously or unconsciously—how the body and mind need to be involved in order to carry out that task. This processing of information about the task gives rise to a strategy with both conscious and unconscious components as to which movements, muscular efforts, and mental processes are to be invested. The task itself—for example, delivering the mail in an urban area, moving the residents in a care home from bed to dining table, or needing to decide on a vital treatment of a patient—exists independently of the person who subsequently carries out the work. The way in which the task is realized and therefore the *loads* that occur in the body are on the other hand unique to the individual person. Each individual has their own way, their "working technique," which depends on both physical and mental factors such as body length, muscle capacity, experience, and attitude. A good working technique is an important precondition for coping with high productivity, while, at the same time, the physical and mental load remains favourable to the body.

The physical and mental load then gives rise to a physiological response, which in a preliminary phase is the immediate, short-term attempt of the body to adapt. For example, sweat is produced when working in a hot environment in order to keep body temperature from rising too much. If the load continues for a longer time, it may be that the adaptation is not sufficient to maintain a physiological balance. Then various more or less adverse effects may arise, for example, an increase in body temperature when heat cannot be completely transferred to the surroundings, or a marked muscle fatigue after repeated physically demanding work cycles. These effects can, in their turn, lead to further physiological responses in a cascading sequence; for example, prolonged heat stress can cause a deterioration of the physical and mental capacity as body temperature gradually rises.

If the original load then drops or is discontinued, the body usually manages to recover to its basic functioning, so that it is ready for new loads. Different bodily functions take different durations for recovery. For example, maximal voluntary muscle strength drops fairly rapidly during heavy muscular work, while it usually recovers within seconds or minutes when the muscle relaxes. On the other hand, it can take hours or days to restore a bodily fluid deficit or nutrition deficiency. The body is constantly switching between being in a stress phase, where it is "being broken down" in what is known as a catabolic process, and being in a recovery phase, where it is once again being "built up" in what is called an anabolic process (see also Chapter 6, Section 6.11). The balance over time between catabolic and anabolic phases is decisive for the well-being of the body and mind. A situation with constantly inadequate recovery after periods of physical or mental stress can lead to long-lasting impaired health, performance, and work ability, in the sense that it will take a very long time to return to full capacity, even if that may be possible. Section 1.4 takes up this phenomenon in relation to the concept of work ability.

Figure 1.1 shows a model of these sequences of events, using the terminology that is often employed within occupational research. The concept of *exposure* means "something the individual encounters." "External exposure" in the model therefore covers both the task that is to be carried out and also the working conditions that

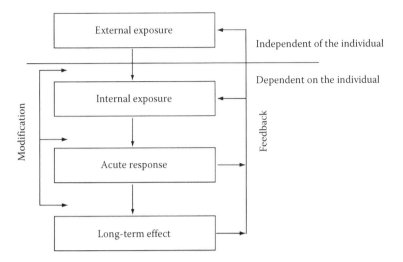

FIGURE 1.1 Model describing the association between the work task (external exposure), the load on the person carrying out the work (internal exposure), and the physiological adaptation of the body to the work in the short term (acute response) and long term (chronic effect). The model also illustrates the fact that the associations are influenced (modified) by who is performing the work, and that both external and internal exposure can be changed through feedback.

affect how that task may be realized, such as technical equipment, working environment, work organization, and psychosocial conditions. The task of "delivering post in postal delivery district Number 14" is therefore an external exposure, just like the number of letterboxes and their positioning, and the time allowances the employer has established for the work. The common denominator for all external exposure is that it is independent of the individual undertaking the work.

The "internal exposure" denotes the loads arising on and in the body when the individual carries out the work. The postal delivery worker will, for example, walk a number of steps, bend down towards the letterboxes, stretch out an arm, and insert the letters. As the development of force is the basic precondition for movement, both of our own bodies and of objects in the world around us, force may be regarded as the primary expression of the internal exposure of the musculoskeletal system. Work postures and work movements are often used as more easily observable expressions of internal exposure. As far as mental loads are concerned, there is no clear counterpart to force as the primary internal exposure.

Both external and internal exposure changes over time. For example, the angle of the upper arm in relation to the vertical changes constantly over a day as the individual moves, and also changes from day to day. To gain an overall picture of exposure, we must understand its amplitude (level), frequency, and duration (Figure 1.2). Amplitude and frequency can be combined in the concept "variation," which represents the change in exposure over time. The variation is thus characterized by how much the exposure changes, how quickly it changes, and whether there is a recurrent pattern of similar exposure elements, such as in repetitive, cyclic assembly work.

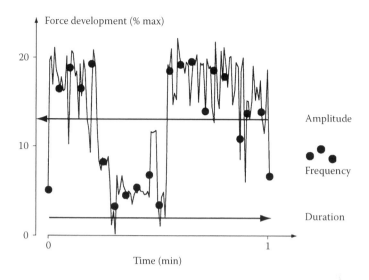

FIGURE 1.2   The three basic dimensions of exposure during 1 min of force development in a muscle: *amplitude*, that is to say, the level of the force, here illustrated by the average force; *frequency*, that is to say how quickly the exposure changes over time, illustrated by the (changes in) force every 3 s of registration; *duration*, that is to say how long the exposure continues, illustrated by the fact that exposure in this case goes on for 1 min.

As described above, internal exposure then gives rise to a cascade of physiological reactions, or "acute responses" (Figure 1.1). These responses are acute in the sense that they return relatively quickly to their initial state if the exposure ceases. Responses that have a longer recovery period or do not recover at all may then be regarded as "long-term effects." Long-term effects can be positive, as an increased physical capacity, following from a training programme, and negative, as muscle pain resulting from prolonged exposure to awkward working postures.

Individual factors influence, or "modify," each of the stages between external exposure and long-term effects. The same task leads to different internal exposures in different individuals, for example, because they choose different working techniques. The same internal exposure produces different physiological responses, primarily due to the fact that individuals differ in their performance capacity. For example, the muscle force required to carry a heavy bag with letters corresponds to 50% of the maximum strength for one individual, but only 20% for another. In the first case, the individual is loaded more and will become tired considerably more quickly than in the second case. There will also be substantial differences between individuals with regard to reaction patterns, even when internal exposures are similar. This applies to both acute responses, for example, drowsiness during night work, and long-term effects, for example, back pain caused by heavy lifting. The model in Figure 1.1 also illustrates the fact that the individual may react to a load by trying to change it, in a so-called "feedback loop." The loop can be involuntary or an act of conscious will. One example of the former is that muscle coordination (i.e., internal exposure) changes with fatigue. An example of voluntary feedback is the attempt to restructure one's work if one has shoulder pain, that is, changing the working technique or reorganizing the time-line of work.

Exposure and response are terms used consistently within physiology and medicine. At the same time as the exposure–response model forms the basis for understanding physiological events, it is applicable as a starting point for the discussion of many issues in working life. If, for example, one wishes to understand the reason why more women than men have neck and shoulder pain, the model helps the user to structure her thinking. Could this be explained by the fact that the external exposure of women and men differ because women have tasks different from those of men? Or is it because women "translate" an external exposure to an internal one in a way different from that of men—for example, because women generally speaking are shorter than men? Or can the same internal exposure give rise to different physiological short- and long-term responses in women and men, for example, because women, generally speaking, have lower muscle strength and oxygen uptake capacity than men? Maybe the differences result from the fact that women experience and report pain and other disorders in a different way than men?

1.3   EXERTION AND FATIGUE

The varying demands of work are thus managed by the individual through varying physical and mental initiatives, leading to varying internal physical and mental loads and physiological responses. Situations that stress the individual significantly in relation to his/her current capacity (ability), and which result in strong physiological reactions, are regarded as *strenuous*.

The degree of exertion can therefore be different in different individuals carrying out the same work. Someone who has a low capacity frequently experiences more exertion when carrying out the same work than someone who has a higher capacity. The working technique is also decisive of what proportion of an individual's capacity he or she uses, and therefore how strenuous the work will be. An individual who only needs to exert half his/her muscle strength or just becomes a little hot and sweaty feels less physically stressed than someone who has to exert maximum strength or becomes very hot and sweaty. On the other hand, when working at one's maximum capacity one is always just as stressed, namely to the maximum, irrespective of capacity. Correspondingly, all healthy people are assumed to be just as (minimally) stressed when completely at rest. A good working technique limits physical and mental exertion and fatigue while maintaining productivity.

Exertion is basically an individual experience. If we wish to acquire a measure of individual exertion, we can therefore in a systematic way ask the individuals to assess the exertion they feel on a scale, for example, the Borg Scale (see Chapter 6, Section 6.12). We can also estimate the degree of exertion from the behaviour of the individual. Breathing heavily, for example, or puffing and blowing and groaning when working, may be a sign of great exertion.

If a job continues for a long time without opportunities for rest, recuperation, and recovery, there is eventually a reduction in the individual's ability to perform, for example, to provide muscle power or to keep mentally alert. This reduction in capacity is physiologically defined as *fatigue*. Physiological fatigue is often accompanied by the individual also feeling tired. Fatigue may, however, be experienced without any visible physiological changes, just as physiological fatigue can exist without any apparent subjective experiences. It is possible to measure physiological fatigue, for example, by the decline in muscle strength, or by certain changes in the electrical muscle activity (EMG—see Chapter 6, Section 6.12). It is also possible to ask the individual to describe and quantify his/her perception of fatigue in a questionnaire, or to carry out tests of vigilance or reaction times. If for physiological or psychological reasons the capacity has declined to such a level that it is no longer sufficient to meet the work demands, the work performance declines both quantitatively and qualitatively. Performance can, therefore, be used as an indicator of fatigue, for example, the number of assemblies made per hour or the extent of errors made during a working day.

The fatigue and sleepiness that result from the diurnal rhythm are not directly related to physical or mental load and have special characteristics (see Chapter 8).

1.4 WORK ABILITY

According to the definition above, fatigue is a transient deficiency in the capacity to carry out work. In normal cases the ability to work is restored relatively quickly to the initial state if one is given the opportunity of recovery. Sometimes the ability to work is, however, limited for a longer time, despite a reasonable recovery period. The concept of *work ability* has come to be used increasingly in daily working life [Nordenfelt 2008]. The concept has usually been used with the approximate meaning of *ability to successfully perform occupational work*. Work ability can be reduced in the longer or shorter term (for hours, days, months, years, or permanently), depending on what has caused the reduction and what remedial measures have been taken. For the individual, the employer, and society, the long-term work ability is often of vital importance.

Work ability depends on the type and level of the physical, mental, and social work demands (see Figure 1.3). Examples of demands of this kind may be: transfer of heavy patients, a long period sitting at a computer, precision work which is visually demanding, high sound levels, shift work, outdoor work at all seasons, high

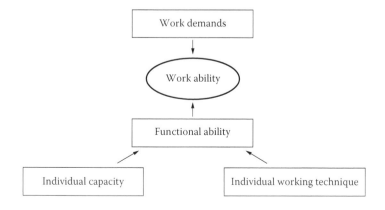

FIGURE 1.3 Factors affecting an individual's work ability.

work pace, rapid decision-making where errors can mean placing other people's lives at risk, reorganizations in which one's job is threatened, solitary work, or dealing with customers' complaints and threats. High demands and demands that are poorly adapted to the individual may lead to inadequate work ability.

The work ability is also dependent on the individual's ability to respond to the various physical, psychological, and social demands of the work, which is to say the individual's functional ability (see Figure 1.3). WHO uses the concepts "functioning and disability" [WHO 2003]. Functioning is, in turn, dependent on individual capacity. The relevant and necessary physical, psychological, and social capacity varies among different professions and situations. Regarding physical capacity, the demands at work normally concern capacity and endurance in developing (muscle-) force, achieving coordinated and precise movements, possessing agility, and sufficient balance, vision, and hearing. In certain cases, therefore, poor training, for example, in strength and endurance, may limit the work ability. There are also a number of psychological and social capacities that may be necessary at work, for example, a good memory, problem-solving skills, verbal ability, stress tolerance, and empathy. If a function of this kind is decisive at work, then a lack of capacity may again lead to insufficient work ability. Several factors can reduce the individual's capacity and functional ability, for example, pain and other physical or psychological problems, disease, complications after accidents, addiction, lack of work motivation, or problematic social conditions. Another cause may be age-related impairment of muscle power, vision, or hearing. The causes of capacity reduction may have arisen in the present job or previous jobs or may be completely unrelated to work. There does not need to be a single cause; it may often be a combination of several, for example, age-related changes in combination with complications after an accident at work.

A further factor which is decisive for work ability is the individual's working technique (see Figure 1.3). The working technique is the individual's way of "translating" work demands (external exposure) into internal exposure (see Figure 1.1). A less suitable working technique may lead to high internal exposures, which may result in loads that are too high relative to the individual's capacity. The work ability may then suffer. An example of this is a care worker who gets back pain because she moves a patient from his/her bed to a wheelchair by lifting the entire weight of the patient instead of making use of transfer techniques resulting in a smaller load. Another example is a computer user who gets a "mouse arm" by working only with the mouse instead of alternatively using keyboard shortcuts. A further example is stress-related disorders caused by completing the day's tasks as fast as possible without taking breaks.

A lack of work ability is, therefore, an expression of an imbalance between the demands of work and the individual functional ability, capacity, and working technique. The same person may have excellent work ability for a specific job, but a worse ability for a different job that makes different demands. For example, an older car assembly worker may find it difficult to keep up with a high work pace on a production line, even if he is used to it; that is to say, his work ability is reduced. If the pace was reduced, the worker would keep up and therefore have adequate work ability. In a corresponding way, the reduction in capacity may have a considerable significance in a particular profession, but little significance in a different profession. For example, a professional singer is presumably not influenced in his/her work ability by injuring a hand. A concert pianist, on the other hand, will suffer a serious reduction in his/her work ability from the same injury.

If an individual's capacity is impaired, the range of jobs suited to their adequate capacity becomes limited. This book contains a large number of examples of the probability of working life, in the longer term, leading to a reduction in physical capacity, caused by, for example, monotonous repeated movements, sustained muscle activity, mental stress, anxiety, or night work.

A common misconception is that the working ability of the individual is equivalent to his/her capacity or functional ability. The two other factors determining work ability are then forgotten—working technique and work demands. The result of such a misconception may be that we try to solve a deficiency in work ability by focusing only on the individual's capacity and miss the two other basic options, which is to say changing the work demands and improving working technique. Chapter 10 presents ideas on how the work ability can be maintained and increased on physiological grounds.

1.5 ADAPTATION FOR GOOD AND FOR BAD

Man—*Homo sapiens*—was at one time shaped by and adapted for life on the African savannah. Life there is something that we can only have vague notions about. Presumably, humans were subjected to a variety of both physical and mental loads. Being able to cope with these loads therefore became vital to survival. This presumably meant being able to see opportunities and risks in different situations, and then if necessary, being able rapidly to react and act. Being able to mobilize all available physical and mental resources was important, for example, to run quickly, develop considerable muscle force, or not be paralysed by a threat. A special asset was man's ability to use tools as a supplement to the very flexible and nimble upper extremities— arms and hands. The sensory organs, the brain, and the musculoskeletal system became optimized on coping with continually changing conditions. Communication, cooperation, and group support were important. The physical and mental adaptability of *Homo sapiens* has been a decisive factor for the success of man in conquering the rest of the globe and for mankind now being on the edge of transcending its limits.

Demands at work are met, as has been described above, by mobilizing physical and mental resources, resulting in various physiological (adaptation) processes. This mobilization of resources can be more or less demanding and protracted. Situations that place great demands on the ability to adapt physically or mentally are sometimes called "stressors." Stressors can be of various kinds. They may, for example, be physical or mental challenges or threats, monotonous repetitive work, work in heat or cold, and work at night. It is, therefore, not merely a question of stress caused by insufficient time or high pace to which people often refer. The more or less conscious experiences of the physical and mental adaptation and the associated physiological reactions are usually called "stress" or "stress reactions." It may be a question of both physical reactions, for example, raised heart rate and blood pressure, and mental reactions, for example, sleep disturbance, anxiety or feelings of insecurity about not being able to cope.

In the short term, human beings are good at managing stressors of this kind, that is to say, exposures requiring adaptation. Adaptability does, however, have its limits. If the stressors last too long, the price of adaptation may be functional disorders and ill-health, for example, cardiovascular disease such as high blood pressure or deficient immune defences, aches and pains, and problems with concentration and memory. Chapter 7 explains further about these so-called allostatic concepts.

Correspondingly, the body adapts to *low* loads. If the bones and joints are not loaded, they weaken. If the muscles are not used, their strength and endurance decrease. If the cardiovascular system is not taxed by occasional hard work, fitness is impaired. Adaptation to low physical load therefore leads to a loss of capacity. If the inactivity continues for a long period, then the risk of ill health from, for example, cardiovascular disease or diabetes increases.

1.6   PRECONDITIONS FOR HUMAN WORK

As explained above, people performing a job have different physical and mental capacities for living up to the demands of the job as regards, for example, muscle strength, precise movements, and memory. People of different ages, for example, vary in maximum muscle strength, endurance and mobility, how quickly and precisely they can select small parts, and how many figures they can keep in their head. In order to achieve a productive and healthy working life, it is important to take into account people's abilities and limitations when designing tasks, workplaces, and tools.

People also have different physical and mental *needs* for their well-being and development. One important need of this kind is to be able to use (activate) and develop their physical and mental functions, that is to say that muscles and joints are allowed to move, and that mental abilities are put to use. Researchers suggest that variation in physical and mental load is a precondition for good functioning and health. Monotonous activity for prolonged periods of time is prejudicial. As a protection against overload and wear and as a stimulus to the development of capacity and functional ability, loads need to vary both as regards level (high–low), complexity (difficult–easy), and type. For example, the joints of the body are dependent on variation in position and force for the joint cartilage to receive an optimum supply of oxygen and nutrients. A fixed joint position over a long period may be directly harmful, as it disturbs this supply. In order to protect against overload and to provide scope for healing and regenerative processes, recurrent periods of physical and mental recovery are necessary. Leisure time and sleep at night are important. Breaks during work are necessary, more frequently so with more demanding physical or mental loads. It is possible to recover exhausted body parts and functions by varying the work so that other body parts and functions bear or share the load. For example, arms and legs are allowed to rest if a task is carried out that mostly involves mental work. Thus, recuperation and recovery can occur even in the absence of a break, pause, or sleep.

A sensible work organization therefore builds variation of this kind into jobs, making variation a natural part of the daily work. An optimal working environment, suitable tools, and equipments make variation possible. Good working techniques utilize these opportunities for variation.

In the long term, it is positive for the individual, business, and society if work takes into account people's physical and mental preconditions and needs and is not merely regarded as a source of income for the individual and a production factor for the company. An exhausted individual, or someone who is in pain or frustrated, cannot be expected to be optimally productive. Work performance in terms of both quantity and quality, may suffer.

Occupational physiology and occupational medicine have a tradition of primarily addressing exposures at work that may result in fatigue, disorders, and other problems. But, as is clear from earlier discussion, it is not harmful to be subjected to load. On the contrary, both physical and mental structures and functions require to be activated so as not to degrade. But inadequate load or overloads of various kinds may also be harmful to health. The challenge in working life is to find those patterns of activity and recovery which in the short term as well as long term promote health and well-being as well as productivity and quality of performance. Sports science and physical education have engaged in similar issues for decades. Today these fields have an advanced knowledge of the effects of different training methods on oxygen uptake, muscle capacity, health, and performance. In comparison with this, the knowledge of the effects of different load patterns in working life is insufficient and vague. We have qualitative knowledge that, for example, "repetitive assembly movements over a long period provide an increased risk of disorders in the lower arm," or that "frequent heavy lifts may result in back pain." However, we lack quantitative knowledge about "how much," "how often," and "how long." One explanation of this lack of knowledge is, among other things, that the patterns of load in working life are far more complicated in time and space than a well-planned training programme for athletes. As is evident in this book, it is therefore only possible to provide general principles for what good work should look like, for example, that it should allow for, and even support, physical and mental variation, or that awkward work postures should only occur to a very limited extent.

1.7 WORK-RELATED MUSCULOSKELETAL DISORDERS

Large parts of this book deal with the musculoskeletal system, that is to say, the skeleton with its joints and ligaments, the muscles with their tendons and bone attachments, and those parts of the nervous system controlling muscle activity. The central role of the musculoskeletal system in the book is justified by the fact that musculoskeletal disorders and injuries have been one of the most widespread health problems in large parts of the industrialized and post-industrial world for several decades now. Several population studies of the prevalence of work-related ill health show that disorders of the musculoskeletal system dominate together with mental conditions [Eurofound 2007; OSHA/EU 2007; SWEA 2008]. The most prevalent musculoskeletal disorders, in turn, are backache and pain in the neck/shoulders or upper extremities, while fatigue and stress lead to the mental outcomes (Table 1.1). About 35% of European workers (male 38%, female 32%) consider their work to have negative effects on their health [Eurofound 2007]. Major differences are,

TABLE 1.1
Percentage of Workers in 27 European Countries Reporting
Different Symptoms Due to Work in 2005

	Percentage
Backache	24.7
Pain in neck, shoulders, or upper extremities	22.8
Fatigue	22.6
Stress	22.3
Headaches	15.5
Irritability	10.5
Injuries	9.7
Sleeping problems	8.7
Anxiety	7.8
Eyesight problems	7.8
Hearing problems	7.2
Skin problems	6.6
Stomach ache	5.8
Breathing difficulties	4.8
Allergies	4.0
Heart disease	2.4
Other	1.6

Source: Eurofound. 2007. *Fourth European Working Conditions Survey.* European
Foundation for the Improvement of Living and Working Conditions.
http://www.eurofound.europa.eu/pubdocs/2006/98/en/2/ef0698en.pdf

Note: More than one symptom may be reported.

however, found between different EU-countries, from 20% to 70%. Major differences are also noted among different work life sectors, with more than 60% of workers in agriculture reporting health problems caused by work, followed by workers in the construction, manufacturing, and health care sectors, where about 40% report problems. Least affected are workers in the financial sector, with about 20% reporting negative health effects of their work. Agricultural workers mostly report physical health problems, whereas workers from the education sector report more mental health problems. The most commonly reported risk factor is repetitive hand and arm movements, reported by more than 60% of all workers, and painful and tiring positions, reported by 45% (Table 1.2).

Musculoskeletal disorder is the most frequently reported cause of work-related disease in many countries (approximately 50% of all cases reported in Sweden) and second after mental problems the most common cause for long-term disability benefits (30% of all cases in Sweden) [Swedish Social Insurance Agency 2008; SWEA 2010]. A rough estimate is that musculoskeletal disorders caused by work cost the Swedish society about 1% of GNP merely in sickness and disability benefits. In addition, disorders cause disruption to the production of companies and organizations and personal suffering. Work has been estimated that approximately 30–40% of all

TABLE 1.2

Percentage of Workers in 27 European Countries Reporting Different Risk Factors during at Least One-Quarter of the Time at Work in 2005

	Percentage
Repetitive hand and arm movements	62
Painful and tiring positions	45
Noise	30
Vibrations	24
Low temperatures	22
Smoke, fumes, dust	18
Chemical substances	14
Infectious material	9
Radiation	4

Source: Eurofound. 2007. *Fourth European Working Conditions Survey.* European Foundation for the Improvement of Living and Working Conditions. http://www.eurofound.europa.eu/pubdocs/2006/98/en/2/ef0698en.pdf

Note: More than one factor may be reported.

ill health in the musculoskeletal system is work related and therefore potentially should be preventable through changes in working life.

Many reported disorders of and injuries to the musculoskeletal system have been regarded by those affected and by health care as work related, even if it has not always been possible to "prove" this by using accepted research methods. When using the term "work-related," it is important to bear in mind that disorders can, indeed, be directly *caused* by the work. Work can also *trigger* or *accelerate* disorders that perhaps have a background in the individual's constitution or degeneration due to age. Without the exposures presented at work, these disorders may well have appeared later in life or may not have manifested themselves at all. Work can also *prevent* or *delay healing* and *rehabilitation* of an injury, which in itself need not be caused or triggered by work. Finally, work may lead to *complications* or in other ways *exacerbate* an injury. In these various ways work may influence the emergence of new cases (incidences) and the occurrence of existing cases (prevalence) of disorders and ill health.

One may ask why work-related disorders of the musculoskeletal system are so common in today's working life, despite the fact that major efforts have been made to prevent and treat them. One important cause among many is that there are a multitude of various risk factors at work for disorders and ill health in the musculoskeletal system: work that is too heavy or too inactive, repetitive operations, badly designed workplaces or tools, vibrations, great precision requirements, a lack of control of one's own work or a lack of support from fellow workers and leaders. The likelihood is great that an individual is faced with one or more such risk factors in today's working life. Synergy between different risk factors increases the risk, for example, if a higher work tempo for a prolonged period is combined with an unsuitable working technique and lack of support from colleagues. It is therefore seldom enough to eliminate single risk factors or improve work using a single measure, for example, ergonomically improved tools. We can compare this with the risk of occurrence of other work-related diseases, for example, poisoning, pneumoconiosis, or hearing damage. In these cases it is often possible to eliminate the risk of ill health by removing a single risk factor, that is to say, the dangerous chemical, the silica dust, or the noise.

Another reason that the efforts, in recent decades, to reduce work-related ill health in the musculoskeletal system have not provided the desired results may be that the expected positive effects have been counteracted by increased rationalization and specialization within working life. This development may, for example, have resulted in less variation in tasks and fewer natural breaks. A classic example is the ergonomic interventions directed at dentists in Sweden during the 1960s. By introducing adjustable chairs in the dentists' clinics, their uncomfortable postures, standing forward-leaning and often twisted, were replaced by a somewhat more comfortable sitting work posture with the equipment within comfortable reach. At the same time, however, the work was rationalized by transferring some of the dentist's former tasks to other professional groups, for example, the receptionist, the dental nurse and the dental hygienist. The work of the dentists became less varied and natural breaks fewer. The reduction in the load level (less leaning forward and fewer twisted postures) was replaced by a much greater duration of sedentary work with small movements in constrained work postures. What is more, a piece rate system was introduced in public dental care, which may have resulted in increased mental stress, and hence increased muscle tension in shoulders and neck. It is not so strange, therefore, that researchers have found that dentists even today have a very high incidence of disorders of the musculoskeletal system, primarily in the neck and shoulders.

Knowledge of occupational physiology is paramount to understanding why disorders and ill health in the musculoskeletal system have become so common, what one could and should do to prevent the problems, and possibly even how work could be designed with the aim of promoting health. The various chapters in this book address the most important requirements in working life from an occupational physiology perspective, explaining in what ways they affect health, well-being, and capacity, and discussing how a healthy working life can be designed by taking these factors into account.

1.8 OCCUPATIONAL PHYSIOLOGY FROM A HISTORICAL PERSPECTIVE

People's physical and mental preconditions and requirements when performing work change only slowly from a historical perspective. Working life today, however, is characterized by constant and rapid technical and organisational change, which can quickly lead to major changes in the work life demands. Some professions disappear through technical developments and because businesses relocate to other countries. New professions appear. The majority of professions in working life remain, however, with more or less radical changes in their contents and technology, and thus in the external exposures presented to the worker.

One example of a profession that has undergone major change since the 1940s is that of forestry work [Attebrant 1995]. In the 1940s and 1950s, work was still carried out manually within many professions, including forestry. Trees were felled, cut into lengths, and trimmed with a handsaw and an axe. The logs were handled manually. This implied heavy physical loads, but it also imposed a limit on productivity. In addition, forestry work was markedly seasonal, as it was primarily carried out during the winter. On the other hand, the work was flexible, and the worker determined to a considerable extent when he wished to work, for how long, and on what tasks. During the 1940s, a shortage of lumberjacks arose, which is why some forestry companies initiated time and motion studies to examine how a more standardized time scheme might contribute to a more efficient use of labour. In connection with these time and motion studies, the trade also initiated studies in work physiology to determine how the work should best be planned to achieve an optimal time-line of exertion and recovery so as to achieve maximally efficient daily work. The study showed that forestry work entailed high-energy metabolism corresponding to an oxygen consumption of approximately 2.5 L/min on average during the working day. This corresponds to approximately 10 times the energy metabolism during rest (see Sections 2.4 and 2.6 in Chapter 2). The total energy metabolism was approximately 21,000 kJ/day (for comparison, the turnover of a female office worker is approximately 9700 kJ/day). The forestry worker's heavy labour may thus have led to a training effect. Measurements during the 1940s and 1950s showed that the forestry workers of those days were very fit. On the other hand, one problem told to be significant among the forestry workers, if not supported by quantitative data, was back pain.

During the 1950s, mechanical aids were introduced in forestry in the form of power saws and barking machines. Productivity increased, but the total energy metabolism of someone working in the trade was the same as in earlier days. Mechanization within forestry continued, and during the 1960s forestry machines were introduced. Many lumberjacks then became machine operators instead. Productivity increased markedly. The poor ergonomic design of the driver's seat, badly placed and stiff control levers, and poor visibility resulted in awkward work postures and high local load on the neck and shoulders. Forestry machine drivers were also extensively exposed to vibrations and shaking. A considerable occurrence of heavy tasks still remained, which provided physical variation from the otherwise sedentary work in the driver's seat. Even if the general workload, measured as energy metabolism, decreased, the average heart rate over a working day was still approximately as high as in completely manual forestry work. The reason was probably that the machine operators were less fit than those lumberjacks carrying out their work entirely manually. The circulatory load in relation to the capacity was therefore still high, as was the load on the back. In addition, local loads increased on the neck and shoulders.

Continued mechanization during the 1970s resulted in a further reduction in general metabolic load. The ergonomic design improved in the new forestry machines. The older, poorly designed forestry machines, however, were still in use, and the incidence of back pain and disorders of the neck and shoulder was still high or even increasing. Within occupational physiology research, the focus shifted from studies of whole-body metabolism to investigations into local muscle loads. The load level (amplitude, cf. Figure 1.1) was of primary interest.

During the 1980s and later, mechanization continued and marked ergonomic improvements were made in order to reduce the local muscle load on the arms, shoulders, and neck. Computers were introduced in forestry machines, which further increased the opportunities of improving both quality and quantity in production. The sophisticated machines and the high demands on productivity resulted in considerable perceptual and cognitive requirements on the operator. The previously heavy and dynamic manual forestry work had now been replaced by sedentary work inside complicated machines, with small repetitive hand/arm movements to control the small multifunctional levers and buttons, as well as exposure to high levels of whole-body vibrations and great mental demands. At the beginning of the 1990s, the sedentary and constrained work in forestry machines comprised 80% of a normal working day in forestry, of which 90% of the time was occupied by repetitive control movements.

A development similar to that in the forestry industry can be seen in many other industries. Technical mechanization has successively reduced the general metabolic load and, in many cases, also the level of the local muscle load on, for example, back and neck/shoulder. This has, however, rarely resulted in the anticipated reduction in the incidence of back, neck, and shoulder disorders. A likely cause is that the occurrence of prolonged sedentary work has increased, sometimes including repetitive arm/hand movements and/or vibrations. As was the case with dentists in the description above, this has led to less variation in work postures and movements and fewer natural breaks. Research from the 1980s and later indicates that there is no acceptable minimal level for prolonged muscle load; even very low levels of load can be a risk if they are sustained without variation or breaks for long periods of time (see Chapter 6).

Thus, the development of occupational physiology can be explained primarily by the need to solve the most obvious problems during different historical periods where there has been an obvious conflict between the demands of work and the physical and mental capacity and needs of people. Consequently, the focus during the period from the 1940s to the 1970s was in general, whole-body physical load, emphasizing energy metabolism, respiration, overall blood circulation, and temperature regulation. During the 1970s and 1980s, the focus moved to the level of local loads on the muscles. From the 1980s onwards, repetitive operations and low-level but prolonged load, attracted more attention, including an increased focus on the time pattern of work and recovery. The effects of mental and psychosocial factors at work on musculoskeletal disorders, particularly in the neck/shoulder region, also received increased attention from the 1980s onwards.

1.9 OCCUPATIONAL PHYSIOLOGY: THIS BOOK

In order to understand the ability of people to carry out their work and the effects that this work may have in the short and long term, we need to understand how the individual functions normally, both physically and mentally, and what bodily reactions occur at different exposures. This is the kind of information offered by the great field of knowledge that is *physiology*, which in its turn can be divided into subcategories, for example, *work physiology*, *muscle physiology*, and *climate physiology*. In this book we bring together the different physiological areas that are relevant to the study of people doing their jobs under the heading *occupational physiology*.

The book concentrates on physiology of the healthy individual and how a lack of balance between demands, capacity, and needs can lead to problems in physiological adaptation. Physiological reactions in extreme work situations, for example, among divers, firemen, or military aircraft pilots are only touched upon by way of exception. The book does neither deals with *sports physiology*, nor with explicit *clinical physiology* such as diagnosis, treatment, and rehabilitation of those with ill health.

With these limitations, the book takes on eight commonly occurring types of work which produce physical and/or mental loads, and therefore has an effect on people's well-being, performance, work ability, and health. The book deals with jobs that entail

- Work demanding high-energy metabolism (Chapter 2)
- Work requiring considerable muscle force (Chapter 3)
- Work in awkward postures (Chapter 4)
- Work with highly repetitive movements (Chapter 5)
- Prolonged, low-intensity, sedentary work (Chapter 6)
- Work with high levels of mental load (Chapter 7)
- Work that disrupts the diurnal rhythm (Chapter 8)
- Work in heat and cold (Chapter 9)

Most professions entail several of these types of exposure at the same time; for example, construction includes elements requiring a high-energy metabolism, a large muscle force, work in extreme postures, repetitive movements, and perhaps also work in heat and cold. Health care may require large muscle forces to be exerted in uncomfortable postures, while the mental loads are also high, and the work is performed in the middle of the night.

The model describing exposure and response (Figure 1.1) forms the basis for the structure in each individual chapter. The chapters begin with a short story of a person with a job typical for the exposures focused by the chapter. Next, some topical questions are posed, which the chapter will answer. The occurrence in working life of the exposures considered is described using statistical data from various countries, for example, the European Agency for Safety and Health at Work and the European Foundation for the Improvement of Living and Working Conditions. The chapters then discuss in greater detail the specific exposure and the normal physiological responses to this. Fact boxes explain and provide greater detail on important points. The chapters also discuss individual factors influencing how the external exposure, the work—is translated into internal exposure—loads on the body—and how the body responds to this internal exposure. The potential health effects of the exposure and the probable mechanisms leading to pain and other problems are dealt with. Each chapter then describes methods for assessing relevant exposures. Suitable interventions are also proposed against problematic working conditions of this kind. The relevant laws and regulations are referred to, and the chapters conclude with a short summary. A selection of key references and "Further reading" tips are to be found after each chapter.

Even if the chapters focus on the physiology of the individual, they also provide information and views on factors at the organizational and societal level determining the working conditions for the individual. For example, several chapters deal with the fact that the allocation of work tasks between individuals in an organization determines the extent of variation in the work of that individual. Also, the laws and regulations presented illustrate the framework that society has established for work.

The chapters may well be read in the order in which they occur, as certain basic sections on energy metabolism, the structure and function of the musculoskeletal system, and certain basic terminology are described in the first chapter in which they are relevant and then referred to in the subsequent chapters. Otherwise, individual chapters can be read separately.

REFERENCES

Attebrant, M. 1995. *Ergonomic Studies of Lever Operations in Forestry Machines.* Master's thesis. Lund: University of Lund.
Eurofound. 2007. *Fourth European Working Conditions Survey.* European Foundation for the Improvement of Living and Working Conditions. http://www.eurofound.europa.eu/pubdocs/2006/98/en/2/ef0698en.pdf
Nordenfelt, L. 2008. *The Concept of Work Ability.* Brussels: P.I.E. Peter Lang.
OSHA/EU. 2007. *European Agency for Safety and Health at Work.* http://osha.europa.eu/topics/msds/facts_html
Swedish Social Insurance Agency. 2008. *Social Insurance in Figures 2008.* http://www.forsakringskassan.se/irj/go/km/docs/fk_publishing/Dokument/Statistik/ohalsostatistik/sfis08_e.pdf
SWEA. 2008. *Work-Related Disorders 2008.* Swedish Work Environment Authority. http://www.av.se/dokument/statistik/officiell_stat/ARBORS2008.pdf
SWEA. 2010. *Occupational Accidents and Work-Related Diseases 2009.* Swedish Work Environment Authority. Report 2010:1. http://www.av.se/dokument/statistik/officiell_stat/STAT2010_01.pdf
WHO. 2003. *International Classification of Functioning, Disability and Health—ICF.* http://www.who.int/classifications/icf/en

2 Work Demanding High Energy Metabolism

Ewa Wigaeus Tornqvist

Photo: Rolf Nyström

CONTENTS

John is 26 years old and has been working for about a year as a bicycle messenger in Stockholm. John chose the job to earn money for a planned trip around the world and to develop his physical capacity before setting out. His work includes delivering various items of mail to different addresses. His tasks during the working day are conveyed to him continuously by two-way radio. John is employed on an hourly basis, but works full time (40 h a week) and he works at piece rate. A typical working day involves cycling between 70 and 100 km and usually delivering 20–40 items. John cycles for ~80% of his working day, that is to say over 6 h a day, while the rest of his working time involves collecting/delivering items of mail and waiting for new assignments. In order to earn as much as possible, John cycles as quickly as he can. He thinks it is hardest to cycle in the centre of the town as he then feels stressed from all the traffic; additionally, he thinks all the exhaust gases and other air pollution he breathes in are annoying. On very hot summer days, the work feels particularly strenuous, and his performance on days like that is lower than normal. When it is very cold and windy in winter, John finds it difficult to dress correctly so that he does not freeze, but will not feel too hot when he is cycling quickly. It is particularly difficult when it is icy and the roads are slippery. John has studded tyres on his bike in the winter, but he has nevertheless fallen down once. Admittedly, that time he got away with just a few scratches, but he thinks it is unpleasant to cycle in traffic when it is icy. Although John has good physical work ability, he is often so tired after work that he does not feel up to any social activities.

2.1   FOCUS OF THE CHAPTER AND DELIMITATIONS VIS-À-VIS OTHER CHAPTERS

Working as a bicycle messenger is heavy physical work characterized by high energy metabolism, and which puts great demands on the body's ability to take up oxygen [Bernmark et al. 2006]. The muscle work is dynamic; that is, the muscles alternate between contracting and relaxing, as in cycling and walking, for example. Heavy dynamic work by large muscle groups makes demands on high energy metabolism and increases the load on the respiratory and circulatory system, which is the focus of this chapter. The load on the circulatory system increases more if the work is carried out under heat exposure, for example, on hot summer days, which is described in Chapter 9. Heavy dynamic work also imposes great demands on muscle force, but the local load on individual regions of the body is usually not as great as in heavy lifting. Work demanding great muscle force is dealt with in Chapter 3. Work demanding great energy metabolism differs appreciably from sedentary work, for example, computer work, both as regards the load on the respiratory and circulatory system and the load on the muscles. Prolonged low-intensity and sedentary work is treated in Chapter 6.

This chapter will take up the following issues:

- What happens in the body during heavy physical work?
- What factors influence the internal load on the body at a certain energy metabolism?
- How can we measure the demands of work on energy metabolism?
- How can we assess the demands of work on energy metabolism in relation to the individual's physical capacity?
- What are the effects of heavy physical work on performance, well-being, and health?
- What exposure limits have been proposed?
- What measures can we take to deal with inappropriately high energy metabolism?
- What do laws, regulations, and ordinances have to say?

2.2   PREVALENCE OF DEMANDS FOR HIGH ENERGY METABOLISM IN WORKING LIFE

Rapid industrial mechanization has reduced the energetic load in many classically heavy industries and sectors such as forestry and agriculture, and the iron and steel industry. The structural transformation of the 1990s also brought with it a reduction in the number of many physically heavy, particularly unqualified, working-class jobs. Despite this, the proportion of all working men in Sweden reporting *heavy physical labour, so that breathing becomes more rapid at least a quarter of the time*, has remained constant at around 20% since the end of the 1980s (own adaptation of Statistics Sweden's Work Environment Surveys 1989–2001) and even increased somewhat to ~24% during the latter half of the 2000s [SWEA 2009]. No corresponding international statistics are available, but the prevalence is probably about the same in other industrial countries. One explanation may be increased rate of work reported both in the statistics and by research. Those occupational groups reporting heavy physical work, according to the above, are to be found primarily within agriculture, forestry, horticulture, and fishing, craft work within the building industry and manufacturing, as well as postal delivery workers (bicycle messengers cannot be distinguished in the statistics). Approximately 13% of all women employed in Sweden report heavy physical work according to the above. The proportion has increased continuously, from ~9% at the end of the 1980s to ~13% during the latter half of the 2000s [SWEA 2009]. The women are primarily to be found within agricultural and horticultural work, nursing and care work, hotel and office cleaning, warehousing, and as postal delivery workers. The increased work rate reported generally in working life is presumably especially obvious within certain female-dominated occupational groups in the public sector, such as nursing and care work, for example, where major staff cuts were made during the 1990s at the same time as the need for care increased rather than decreased. These changes presumably help explain the fact that the proportion of women reporting heavy work has increased since the end of the 1980s. Heavy work within many female-dominated service occupations may be expected to continue to increase, as the need for nursing and care of the elderly is increasing because of demographic change. The problem of heavy physical work is also exacerbated by the fact that employees are becoming older, as their physical capacity declines with increasing age. The proportion who are 55 years of age and older has increased by 5% points since 1990, and now comprises one-fifth of working people in Sweden. The proportion of older employees is expected to increase for some years until the large population groups born in the 1940s retire.

The results of the work environment surveys mentioned above agree reasonably well with the results in a Swedish sample of the general population, where ~24% of the men and 11% of the women were assessed as having high energetic load [Wigaeus Tornqvist et al. 2001].

2.3 ENERGY METABOLISM AND PHYSIOLOGICAL ADAPTATION WITH INCREASED ENERGY DEMAND

The energy requirement varies greatly between different occupations (Figure 2.1). In light physical work, for example, sedentary computer work, the energy metabolism increases during the work to approximately twice the basal metabolic rate, while in very physically demanding jobs, for example, a bicycle messenger, the energy metabolism may increase to six times the basal metabolic rate. For a man weighing 75 kg, the basal metabolic rate (seated at rest) is ~7560 kJ/day (see also Section 2.4).

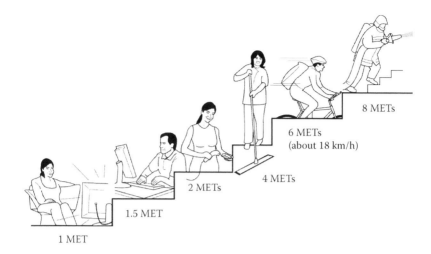

FIGURE 2.1 Energy metabolism at rest and for different tasks, expressed as multiples of basal metabolic rate, METs. One MET corresponds to the energy metabolism seated at rest (4.2 kJ/kg of body weight and hour). (Adapted from Ainsworth BE. et al. 1993. *Med. Sci. Sports Exerc.* 25(1):71–80.) Illustration: Niklas Hofvander.

The basal metabolic rate per kilogram of body weight is somewhat lower in women compared with men, because women have a greater proportion of fatty tissue, whose metabolic activity is low. Basal metabolism also varies between individuals of the same gender and a basal metabolic rate of 10% above or below the mean value is not unusual. These individual differences may explain why certain people can remain slim, while others increase in weight, even if they both eat the same amount and are equally physically active.

Human beings are adapted to movement and physical activity. Physical activity demands energy, and the skeletal muscles are unique compared with all the other tissues of the body in their ability to increase energy metabolism. In extremely physically demanding activities, the energy metabolism normally increases to 10–20 times the basal metabolic rate; for elite athletes with extremely high physical capacity, it can be up to 30 times the basal metabolic rate. The difference between different individuals' energy metabolism, therefore, results mainly from the degree of muscle work (Figure 2.1).

The ability to perform physical work is dependent on the ability of the muscles to convert chemically bound energy in the food into mechanical energy in the form of muscle work. In heavy dynamic muscle work in which large muscle groups are involved, oxygen uptake ability is of great significance for the release of enough energy for muscle work. Work of this kind results in a load on respiratory and circulatory systems. The internal load on the individual depends on how great a proportion of the individual's maximum capacity is used in the work. In healthy individuals, it is primarily the size of the heart's stroke volume that determines the individual's maximal oxygen uptake capacity (VO_2 max L/min, where V stands for the volume and O_2 is the chemical symbol for oxygen). VO_2 max by definition corresponds to the maximum volume of oxygen uptake per unit of time that can be measured in an individual. This ability varies considerably between different individuals, depending on individual factors such as heredity, age, gender, and state of health, as well as lifestyle factors such as physical training. In maximum work a fit 30-year-old man with "good" genes (as regards physical capacity) can increase his energy metabolism ~20 times compared with basal metabolic rate, while the corresponding increase for an unfit 50-year-old woman with less "good" genes is 5–10 times.

2.4   ENERGY METABOLISM

The nutrients that the body needs consist of carbohydrates, fats, proteins, minerals, vitamins, and water. These substances have to be supplied to (1) cover the body's energy needs, (2) cover the need for substances to build various tissues, (3) cover the need for substances to build up enzymes, hormones, and other important substances, and (4) replace the body's losses of important substances. Of these nutrients, it is only carbohydrates, fats, and proteins whose energy content can be converted from chemically bound energy into mechanical energy for muscle work. Carbohydrates exist in the form of glucose in the blood and are stored in the form of glycogen in the liver and muscles. Proteins are used primarily for construction and renovation work in the body. The chemically bound energy that is converted, but not used for mechanical energy for muscle work, is converted into heat (see Section 2.5).

The energy metabolism increases linearly with increased work load. Under normal circumstances, presupposing adequate energy supply, carbohydrates, and fats are (mainly) oxidized with the help of oxygen into carbon dioxide and water, the so-called *aerobic* oxidation. The chemically bound energy in the nutrients is converted and used partly to form adenosine triphosphate (ATP). ATP is an energy-rich substance which, when it breaks down, can directly transfer its energy to mechanical work in a muscle contraction. Under circumstances when the oxygen supply is not sufficient for an *aerobic* oxidation, the conversion of energy occurs *anaerobically*, and then the carbohydrates are converted without the presence of oxygen, whereby lactic acid is formed (Figure 2.2).

Anaerobic metabolism can only contribute energy for a very short period, which is why the body is dependent on *aerobic* metabolism to maintain the necessary ATP level. The supply of energy for muscle work lasting for more than a minute or so is therefore dependent on oxygen supply and the supply of fat and carbohydrates, as well as necessary *enzymes* (a substance needed for a certain chemical reaction to take place).

In *anaerobic* metabolism, energy is released very quickly, but the stock is very limited. The speed with which energy is released in *aerobic* oxidation of carbohydrates, and particularly fats, is considerably slower, but the depots are considerably greater, particularly those of fat. There is an inverse proportionality between the size of the energy stock and the speed whereby the energy is released (Table 2.1).

The energy metabolism per day at rest is ~6000 kJ for a woman weighing 60 kg and 7600 kJ for a man weighing 75 kg (see also Fact Box 2.1). In physically light occupations the energy metabolism is ~9200 kJ for a woman and 12,400 kJ for a man, while a moderately heavy job requires ~10,500 and 15,300 kJ/day, respectively.

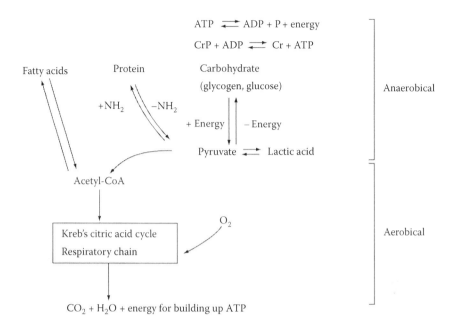

FIGURE 2.2 Build-up of ATP in a cell. In the fission of ATP, energy is released, which is transferred directly, for example, to mechanical work in a muscle cell. The energy released in other types of metabolism is used to build up ATP. The metabolism of ATP and creatine phosphate (CrP) and of carbohydrate to lactic acid, via pyruvate, occurs without oxygen, that is, to say anaerobically. Fatty acids and pyruvate (from carbohydrate and protein) are metabolized with the help of oxygen—that is, aerobically—in Krebs' citric acid cycle. (Adapted from Åstrand I. 1990. *Arbetsfysiologi.* 4th ed. Stockholm: Norstedts förlag (in Swedish), p. 16. With permission.)

TABLE 2.1
Energy Sources for Muscular Work

Energy Source	Body's Store (kJ)	Time (min)	Relative Speed
ATP, CrP[a] in muscles and anaerobic metabolism of carbohydrates	80	1	1.0
Oxidation of blood glucose	320	4	0.5
Oxidation of liver glycogen	1500	18	0.5
Oxidation of muscle glycogen	6000	70	0.5
Oxidation of fat	33,700	4018	0.25

Source: Data from Jones DA., Round JM. 1990. *Skeletal Muscle in Health and Disease.* Manchester and New York, NY: Manchester University Press.

Note: The values for the total size of energy storage (the body's store) in kJ, the time the energy source last would last as a sole energy source in work corresponding to 80% of maximum aerobic capacity and the relative speed for the release of energy.

[a] Creatine phosphate, which is used for building up ATP from adenosine diphosphate, see Figure 2.2.

FACT BOX 2.1

Basal metabolic rate

- Energy metabolism is often expressed in multiples of the basal metabolic rate, so-called metabolic units or METs.
- One MET corresponds to the energy metabolism seated at rest.
- Seated at rest ~4.2 kJ (1 kcal) per kg of body weight is expended per hour (4.2 kJ × kg^{-1} × h^{-1}).

The relative proportion of carbohydrate and fat metabolism, respectively, depends primarily on the following factors:

1. Type of work: intensity and duration
2. State of fitness: *fit* or *unfit* individuals
3. Diet: carbohydrate-rich or carbohydrate-poor diet

2.4.1 TYPE OF WORK

At rest, or during shorter periods of light-to-moderately heavy physical work, carbohydrates and fat contribute approximately equally to the energy supply (presupposing that the individual eats a normal varied diet) (Figure 2.3). When the work load increases, a gradual increase occurs in the proportion of carbohydrates metabolized, which is effective from an energy viewpoint, as the amount of energy released in the metabolism of 1 L of oxygen is higher in carbohydrate metabolism than in fat metabolism. In very heavy, almost maximal, work it is carbohydrates that represent 100% of the energy supply (Figure 2.3). At such a high level of workload the oxygen supply is insufficient, and the carbohydrates are partly metabolized *anaerobically*, whereby there is an accumulation of lactic acid, which means that the functional capacity of the muscle cells deteriorates.

The longer the work continues, the higher the proportion of fat metabolized. In moderately heavy work, which can go on for 4–6 h (including rest breaks, but without the supply of nutrients), fat can contribute up to 60–70% of energy metabolism. Although carbohydrate metabolism declines in prolonged moderately heavy work, the supply of carbohydrates may limit endurance as a result of limited stocks. In work corresponding to 75% of the individual's maximal aerobic capacity, the glycogen level in the muscles drops successively and reaches zero after one-and-a-half hours of work. The more glycogen is available from the beginning, the longer the work can continue. The glycogen levels can be increased by consuming carbohydrate-rich food. By initially emptying the glycogen stores through eating carbohydrate-poor food, the depots are stimulated to restock with extra glycogen.

2.4.2 STATE OF FITNESS

The ability to make use of fat as a fuel depends on the capacity for oxygen transport. The proportion of fat metabolism at a particular work load depends on the proportion

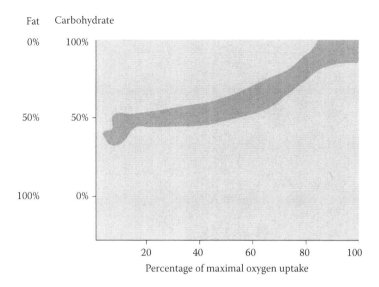

Fat   Carbohydrate

FIGURE 2.3   The proportion of fat and carbohydrate metabolism at different work loads expressed as a percentage of the subject's maximal oxygen uptake. The proportion of carbohydrate metabolism increases with increasing work load after ~50% of maximal oxygen uptake. (Adapted from Åstrand PO. et al. 2003. *Textbook of Work Physiology. Physiological Bases of Exercise.* 4th ed. Champaign, IL: Human Kinetics, p. 373. With permission.)

of the maximal aerobic capacity used (Figure 2.3). As physical exercise can increase the individual's maximal aerobic capacity (VO_2 max/min), the ability to make use of fat as an energy source at a particular oxygen uptake also increases. In prolonged work, it is a great advantage to be able to use more fat for oxidation, as the fat stores are definitely greater than the carbohydrate stores (Table 2.1). Furthermore, as fit individuals use a smaller proportion of their maximal aerobic capacity in work at a particular work load, the production of lactic acid is lower in fit people compared with that in unfit people.

2.4.3   DIET

Adaptation to a fat-rich and carbohydrate-poor diet results in lower glycogen levels in the muscles and liver, and this results in an increased fat metabolism during work to save glycogen.

2.5   MECHANICAL EFFICIENCY

Some of the chemically bound energy converted in metabolizing various substances is turned into heat. This heat production means that the human body temperature can be kept relatively constant at 37°C, even though the ambient temperature is often lower. In muscle work, relatively large amounts of heat are produced, and during dynamic work by large muscle groups 70–75% of the energy

TABLE 2.2
Maximal Efficiency in Different Physical Operations

Activity	Efficiency (%)
Walking uphill on a 5° slope, without load	30
Walking on a level surface, without load	27
Cycling	25
Going up and down the stairs, without load	23
Using a heavy hammer	15
Lifting weights	9
Shovelling in an upright posture	6
Using a screwdriver	5
Shovelling in stooped posture	3

Source: Data from Kroemer KHE., Grandjean E. 1997. *Fitting the Task to the Human. A Textbook of Occupational Ergonomics.* 5th ed. London: Taylor & Francis.

metabolism is converted into heat. The mechanical efficiency of the human being—that is, the proportion of the total energy metabolism used to carry out external work—therefore reaches a maximum of 30%. Usually, however, the mechanical efficiency is lower. In work by small muscle groups, primarily work with the arms, and elements of static working operations, for example, in work with non-neutral work postures, the mechanical efficiency is considerably lower (Table 2.2).

The heavier the work, the more heat formed, which must be emitted so that the body does not overheat. Blood circulation to the skin increases in order to increase heat transport from the working muscles out to the surface of the body. The secretion of sweat increases in order to increase heat transfer (see also Sections 2.8 and 2.9.2 as well as Chapter 9). Every litre of sweat that evaporates from the skin means a heat transfer corresponding to 2450 kJ. Although heat transfer increases with physical work, the core body temperature also increases. Normally, core body temperature is dependent on the relative work load, that is, the proportion of the individual's VO_2 max that is used. The increase in body temperature is considered to be an active regulator for creating a gradient for heat flow from the core of the body to the skin and to stimulate sweating. In work at a particular work load, the body temperature increases continuously for ~40–50 min, and then levels off at a level dependent on the relative workload. The correlation between body temperature (T) and relative load (VO_2/VO_2 max) can be described using the formula

$$T = 36.5 + (3.0 \times VO_2/VO_2 \text{ max})$$

In John's job as a bicycle messenger, he uses on average 30% of VO_2 max (see Section 2.11.1) and reaches a body temperature of ~37.4°C.

2.6   OXYGEN UPTAKE

For practical use it is difficult to measure energy metabolism directly, as this requires measuring the amounts of carbohydrate, fat, and protein supplied and used respectively, how great the body's energy stores are, and how they have changed. Oxygen uptake, like energy metabolism, increases linearly with increasing work load (Figure 2.4). Energy metabolism can therefore be measured indirectly on the basis of what is known as the *energy coefficient* by measuring oxygen uptake. The *energy coefficient* is the amount of energy metabolized when 1 L of oxygen is used. In practical use we use an energy coefficient of 20 kJ (20.2 kJ at rest and 20.6 kJ at work).

The volume of oxygen (VO_2) used in an average adult while seated at rest corresponds to ~3.5 mL/kg of body weight per minute (3.5 mL $O_2 \times kg^{-1} \times min^{-1}$). A person weighing 75 kg therefore uses ~0.26 L of oxygen per minute at rest, which equals an energy metabolism of ~5.25 kJ/min or 7560 kJ/day. While oxygen uptake for maximal work (VO_2 max/min) in a young, fit man can increase to perhaps 5 L/min, that is to say ~20 times the basal metabolic rate, the corresponding value for an older unfit woman can be 1.5 L/min. For a woman weighing 75 kg, this means barely a sixfold increase on basal metabolic rate, but if she weighs, for example, 52 kg this means that she can increase her basal metabolic rate of ~0.18 L/min by a good eight times.

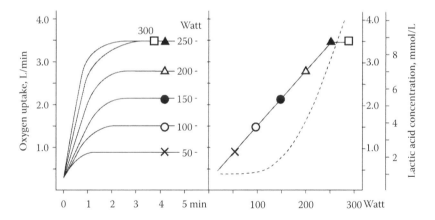

FIGURE 2.4   Oxygen uptake at submaximal and maximal levels in an individual. On the left oxygen uptake is illustrated at varying work loads in relation to time; on the right, oxygen uptake measured every 4–5 min and the corresponding lactic acid concentrations in the blood are shown. Solid line = oxygen uptake; broken line = lactic acid concentration. Oxygen uptake at 50 W is ~0.9 L/min, and at 100 W is ~1.5 L/min. This individual's maximal oxygen uptake is reached at an intensity of 250 W. Note that the lactic acid concentration in the blood begins to increase markedly after 50% of the aerobic capacity has been reached. Lactic acid formation derives primarily from the beginning of the work. (Adapted from Åstrand PO., Rodahl K. 1986. *Textbook of Work Physiology.* 3rd ed. New York, NY: McGraw-Hill Book Co., p. 300. With permission.)

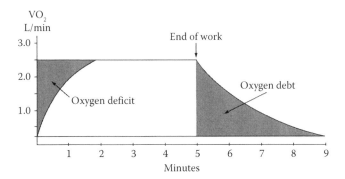

FIGURE 2.5 Oxygen uptake (VO$_2$ L/min) increases at the beginning of moderately heavy work until a level is reached where the uptake corresponds to the tissues' need for oxygen, what is called steady state. During the first minutes therefore an oxygen deficit arises. After the end of the work oxygen uptake drops slowly, the oxygen debt is repaid. Basal metabolism is ~0.25 L/min. (Adapted from Åstrand I. 1990. *Arbetsfysiologi.* 4th ed. Stockholm: Norstedts förlag (in Swedish), p. 38. With permission.)

At the beginning of dynamic physical work the oxygen uptake increases until what is called a *steady state* is possibly reached, when the uptake of oxygen corresponds to the demands of the muscles (Figure 2.5). It takes several minutes to adapt breathing and circulation from the basal metabolic rate to the increased demands of physical work. During this so-called *oxygen deficit* the demands of the muscles for oxygen are not being met by the oxygen supplied through circulation. In light physical work, the amount of oxygen that is stored in the myoglobin is sufficient for energy metabolism to occur *aerobically*, despite the fact that the oxygen supply through the circulation has not had time to adapt to the increased demands. With somewhat heavier work, the oxygen in the myoglobin is insufficient, and the energy is provided partly *anaerobically*, whereupon lactic acid is produced. The heavier the work, the more lactic acid is produced at the beginning of the work (Figure 2.5). If we want to find out how much oxygen a particular job demands, the work has to continue for at least 3–4 min so that we can be sure that we have reached this *steady state*.

At a higher work load the oxygen uptake increases, and it levels out at a higher *steady-state* level (Figure 2.4). When the work load becomes very high, the amount of oxygen taken up is insufficient for the work to be performed with *aerobic* energy supply, and precisely as at the beginning of the work the energy supply is partially *anaerobic*. The higher the load, the more *anaerobic* the energy supply and the higher the lactic acid concentration in the working muscles and in the blood (Figures 2.4 and 2.6). The lactic acid concentration in the blood begins to increase appreciably after ~40–50% of maximal aerobic capacity has been reached (Figure 2.6). If the work continues at this higher load, lactic acid accumulates in the muscles and blood, and the stiffness, fatigue, or aches then felt in the working muscles probably result from the lower pH value. The slow decline in oxygen uptake after the end of work results from the fact that the so-called *oxygen debt*

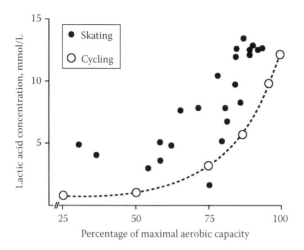

FIGURE 2.6 Lactic acid concentration in the blood at different loads expressed as a percentage of maximal oxygen uptake during cycling and speed skating. Note the higher lactic acid concentration in skating compared with cycling. The static load on the thigh muscles resulting from the "sitting work posture" in speed skating results in poorer efficiency compared with dynamic cycling. (Adapted from Ekblom B., Hermansen L., Saltin B. 1967. *Hastighetsåkning på skridsko*. Idrottsfysiologi, Rapport no 5. Stockholm: Framtiden. With permission.)

from the beginning of the work, and possibly during the work, has to be paid back (Figure 2.5).

Compared with dynamic work—that is, alternation between contraction and relaxation—the blood circulation in static muscle contractions (see Chapter 6, Section 6.6) is often impaired and oxygen supply can therefore be insufficient, whereupon the *anaerobic* supply of energy increases and lactic acid is formed. Impaired blood circulation also results in impaired removal of metabolites, which further influences the accumulation of lactic acid, for example (cf. the lactic acid concentration in the blood in cycling and in skating, respectively, in Figure 2.6).

2.7 PULMONARY VENTILATION

In the transition from rest to physical work, pulmonary ventilation (litres of air per minute) increases to accommodate the increased oxygen requirement in the blood and to vent the excess of carbon dioxide. In light-to-moderately heavy physical work, pulmonary ventilation increases linearly with an increase in oxygen uptake (Figure 2.7). After this, pulmonary ventilation increases more per litre of oxygen taken up as we get closer to the individual's maximal aerobic capacity. This so-called *hyperventilation* is caused by acidic products, for example, lactic acid, from energy metabolism. The degree of acidity in the blood stimulates the respiratory centre to increase pulmonary ventilation to vent the excess carbon dioxide.

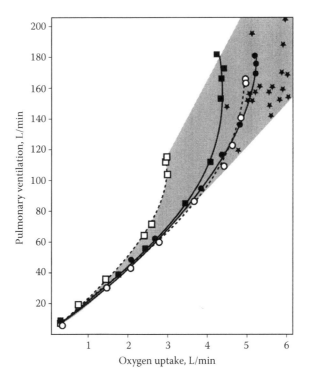

FIGURE 2.7   Pulmonary ventilation increases with an increase in oxygen uptake. The lines correspond to the values for four different people in dynamic work with large muscle groups (cycling or running). The stars (*) represent individual values from elite sportsmen in connection with the determination of maximal aerobic capacity. Most individuals with a maximal oxygen uptake higher than 3.0 L/min lie within the shaded area. (Adapted from Åstrand P-O. et al. 2003. *Textbook of Work Physiology. Physiological Bases of Exercise.* 4th ed. Champaign, IL: Human Kinetics, p. 190. With permission.)

When John is exposed to air pollution on streets with dense traffic, the uptake of pollutants into his body increases (*internal exposure*) when he cycles fast, because pulmonary ventilation and blood circulation to the lungs increase (see also Section 2.8 and Fact Box 2.2).

FACT BOX 2.2

The body's physiological adaptation to higher energy demands can increase the risk of negative effects of exposure to different kinds of air pollution.

If someone works in an environment with air pollutants, their body is subjected to a greater *internal exposure* to pollutants during physical work compared with being at rest. The greater pulmonary ventilation increases the amount of pollutants inhaled, and if the substance dissolves in the blood it is distributed through the vascular system to various organs in a greater quantity per unit of time during work, compared with at rest. It is important to bear this in mind in connection with a discussion of the effects of exposure to air pollutants that exist at workplaces or generally in the surrounding environment. For substances easily absorbed into the blood and fatty tissues, such as the brain, for example, the risks of central nervous system effects, (e.g., slower reaction times in exposure to solvents) are greater during physical work than at rest. Another organ that is sensitive to pollutants is the liver, which functions as the body's treatment plant by breaking down pollutants into metabolites which can be excreted from the body. The impaired liver perfusion that occurs in heavy physical work, as compared to rest, can result in an impairment of the metabolism and excretion of various pollutants.

2.8   BLOOD CIRCULATION

In the transition from rest to heavy physical work, oxygen uptake—as has previously been mentioned—may increase up to 30 times at the same time as the cardiac output; that is, the total blood circulation (heart rate × stroke volume) increases fivefold. The greater increase in oxygen uptake compared with the cardiac output is due to a greater proportion of the oxygen content of the blood being used in heavy work; that is, the oxygen content in mixed venous blood decreases, and the so-called arteriovenous oxygen difference increases (Figure 2.8). According to Fick's principle,

Oxygen uptake (VO_2) = cardiac output (Q)
$$\times \text{arteriovenous oxygen difference } [(a - \bar{v})O_2 \text{ diff}]$$

The cardiac output increases linearly with an increase in oxygen uptake, from ~5 L/min at rest to ~25 L/min at 70–80% of maximal aerobic capacity (Figure 2.8). The increase is achieved partly through the heart rate in most cases increasing linearly with greater oxygen uptake, and partly through the stroke volume increasing on a rising curve, reaching its maximum value at 40–50% of the maximum aerobic capacity. At rest the heart rate is normally ~60 beats/min and the stroke volume is ~80 mL, increasing to ~200 beats/min and 125 mL, respectively, at maximum work. Individual variations, however, are great, which is described below.

The increase in the arteriovenous oxygen difference results partly from an actual increase in oxygen use per volume unit of blood in the working muscles, and partly from a redistribution of the circulating blood volume, so that a much larger proportion of the blood supply goes to the working muscles (Figure 2.9). The higher the haemoglobin content in the blood, the higher the oxygen content in the arterial blood. Women have on average a lower haemoglobin content compared with men (on average 13.9 g and 15.8 g/100 mL of blood, respectively), which results in a lower arterial oxygen content in women—~16 mL compared with ~19 mL of oxygen/100 mL of blood, and thus a lower arteriovenous oxygen difference, which presumably explains the moderately higher cardiac output in women at a particular oxygen uptake. At rest the oxygen content in mixed venous blood is ~10–12 mL/100 mL of blood dropping

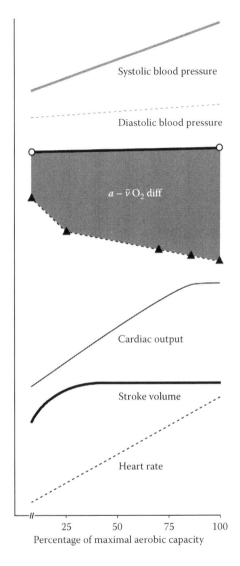

FIGURE 2.8   Heart rate, cardiac output, arteriovenous oxygen difference, and systolic blood pressure increase with a rise in oxygen uptake or work load. The stroke volume normally reaches maximum value at ~40–50% of maximal aerobic capacity. O = arterial oxygen concentration (*a*), ▲ = oxygen concentration in mixed venous blood ($\bar{v}$). (Adapted from Åstrand I. 1990. *Arbetsfysiologi.* 4th ed. Stockholm: Norstedts förlag (in Swedish), p. 71.)

to ~2 mL/100 mL of blood in very heavy work, while the oxygen content of venous blood from a hard-working muscle can be zero.

Out of the cardiac output at rest of ~5 L/min, about 80–85% is distributed to the internal organs, and 15–20% to the skin and muscles (Figure 2.9). During heavy work, the figures are reversed: 80–85% of the cardiac output of ~25 L/min is directed to the skin and muscles, while 15–20% goes to the internal organs. In absolute figures,

Organ/tissue	Blood flow at rest	Blood flow during heavy physical work
Blood from the heart (cardiac output)	5 L/min 100%	25 L/min 100%
Coronary vessels	0.2 L/min 4–5%	1 L/min 4–5%
Gastrointestinal tract	1.25 L/min 25–30%	0.75 L/min 3–5%
Kidneys	1 L/min 20–25%	0.5 L/min 2–3%
Central nervous system	0.75 L/min 15%	1.25 L/min 4–6%
Skeleton	0.15 L/min 3–5%	0.125 L/min 0.5–1%
Skin	0.25 L/min 5%	
Muscles	0.75 L/min 15–20%	20 L/min 80–85%

Cardiac output 5 L/min	Rest L/min	Organ/tissue	Heavy work L/min	Cardiac output 25 L/min
	0.2	Coronary vessels	1	
	1.25	Gastrointestinal tract	0.75	
	1	Kidneys	0.5	
	0.75	Central nervous system	1.25	
	0.15	Skeleton	0.125	
	0.25	Skin	} 20	
	0.75	Muscles		

FIGURE 2.9 Cardiac output can increase by a factor of five in the transition from rest to very heavy physical work, and at the same time the relative distribution of blood flow to various organs changes. Notice the marked increase in blood flow to the skin and muscles in heavy work. The blood flow to the vessels of the heart itself and to the central nervous system also increases in heavy physical work. (Adapted from Åstrand I. 1990. *Arbetsfysiologi*. 4th ed. Stockholm: Norstedts förlag (in Swedish), p. 70.)

this means a 20-fold increase in blood flow to skin and muscles, from ~1 L/min at rest to 20 L/min during heavy work. The blood flow to various organs is regulated by changes in the diameter of the smallest arteries (the arterioles). In increased muscle work, when the need for oxygen to the working muscles increases, the vessels dilate (vasodilation) while they contract (vasoconstriction) when there is less need for oxygen. Apart from skin and muscles, the lungs also of course receive greatly increased blood circulation, but in absolute figures the coronary vessels in the heart and the central nervous system also receive greater blood circulation. The stomach and intestines, liver and kidneys can, on the other hand, receive a decreased blood supply during heavy work. In the transition from rest to physical work one can sometimes get what is called a "stitch," which presumably results from the reduced circulation to the stomach and intestinal system.

The heart muscle pumps blood out into the vascular system at a certain pressure. When the left ventricle contracts, blood is pumped out through the aorta, and the pressure in the aorta is usually ~120 mm Hg (16.0 kPa), which is called the systolic blood pressure. When the aortic valves then close between heartbeats, the blood pressure in the aorta drops, and the lowest pressure that can be measured in the aorta is usually ~80 mm Hg (10.6 kPa), which is called the diastolic blood pressure. The normal pressure in young healthy people at rest is ~120/80 mm Hg.

In connection with physical work, blood flow to the working muscles increases in order to meet the demand for oxygen supply and to remove carbon dioxide and other metabolites, for example, lactic acid, and excess heat. For this it is necessary for the heart muscle's contractive force, as well as the cardiac output and blood pressure, to increase. Systolic blood pressure increases more than diastolic pressure in physical work (Figure 2.8). The blood pressure reaction during work can provide important information about the individual's cardiovascular system. The blood pressure increase during work is, for example, greater in older individuals compared with younger ones. This is presumably the result of a decreased elasticity in the vessels with increasing age. In, for example, cycling at a heart rate of ~150 beats/min, healthy men of 20–30 years of age have a pressure of ~150/80 mm Hg, while blood pressure in men between 50 and 60 years of age is ~210/95. The individual variation in blood pressure reaction at an increased work load is, however, relatively large.

2.9   FACTORS AFFECTING INDIVIDUAL LOAD AT A PARTICULAR EXTERNAL EXPOSURE

Load on the individual, the *internal exposure* at a certain *external exposure*—for example, cycling 100 km and delivering 30 items of mail during an 8-h working day—is affected by a number of factors. The internal load results from, among other things, factors at work, for instance, how certain operations are carried out, how the work is organized, for example, as regards breaks, and what equipment is used. The internal load is also affected by factors in the environment, such as how hot it is. What is more, there are individual characteristics, for instance age, and lifestyle factors, such as physical exercise, which are of great significance for individual load at a certain external exposure. Psychological factors are not dealt with in this chapter, but these aspects are taken up in Chapter 7.

2.9.1   FACTORS AT WORK

2.9.1.1   Dynamic and Static Muscle Work

Seen from a functional perspective, two different types of muscle work are distinguished, *dynamic* and *static* muscle contractions, respectively (see Chapter 6, Section 6.6).

Most operations contain both dynamic and static work. Work with the arms and hands lifted—for example, carpentry, painting, and cleaning work—involves static loads primarily in the neck and shoulder muscles. The efficiency and the maximal performance (VO_2 max/min) declines appreciably with elements of static work operations (see Table 2.2). In hammering nails at different heights, for example, oxygen uptake is relatively constant if the nails are being hammered in with lowered or lifted arms, but the performance, the number of nails/min, is considerably lower when nailing into a ceiling compared with nailing at bench height (Figure 2.10). The efficiency when nailing with lifted arms is therefore lower. Heart rate, blood pressure, and lactic acid concentration are higher when nailing into a ceiling compared with nailing at bench height.

2.9.1.2   Work by Small Muscle Groups

As mentioned earlier, the body's ability to achieve maximum oxygen uptake is only reached in dynamic muscle work by large muscle groups. In maximum work with the arms, approximately only 70% of VO_2 max is achieved. If leg work resulting in maximum oxygen uptake is distributed in both the arm and leg muscles, there is often no noticeable increase in VO_2 max, as the maximal uptake capacity can be achieved with legwork. On the other hand, the work can be carried out for a longer period, as the load is distributed over a greater muscle mass.

In submaximal work by small muscle groups, compared with large groups, the cardiac output at a certain oxygen uptake is largely the same, but the heart rate is higher and its stroke volume is lower. The blood pressure reaction is significantly higher at a specific cardiac output. The higher blood pressure and higher heart rate in work by small muscle groups result in poorer efficiency compared with work by large muscle groups. The reason for the higher blood pressure is presumably a vascular contraction (vasoconstriction) in the inactive muscles to offset a vascular dilation (vasodilation) in the muscle that is working.

2.9.1.3   Breaks

Extremely heavy operations in which the employee can choose their own work rate can be made easier, without any decline in production, by inserting short breaks. Figure 2.11 shows the heart rate of a person carrying pieces of iron weighing 30 kg from a machine to a palette. When he, for example, carried 14 pieces in a row and subsequently took a break, continued to carry 14 pieces and took another break, he reached a heart rate of ~150 beats/min at the end of each work cycle. The length of the break was always 1.5 times longer than the duration of each work period. When he reduced the number of pieces of iron before he took a break—still with a break time 1.5 times longer than the work period—his heart rate dropped. When he took four pieces of iron each time, and therefore took 14 short breaks, he reached a heart rate of 110 beats/min. The shorter the period of work, the lower the heart rate, despite the fact that the production was the same. After ~13 min he had carried 56 pieces of iron in each case. By further reducing the number of pieces of iron the heart rate dropped further, but the work seemed disjointed.

Similar results have also been shown in laboratory trials with cycling and running as types of work. The load, measured both as heart rate and concentration of lactic acid in the blood, declined further the shorter the working period. In these trials the effect of the length of the break was also studied. Interestingly enough, there was no

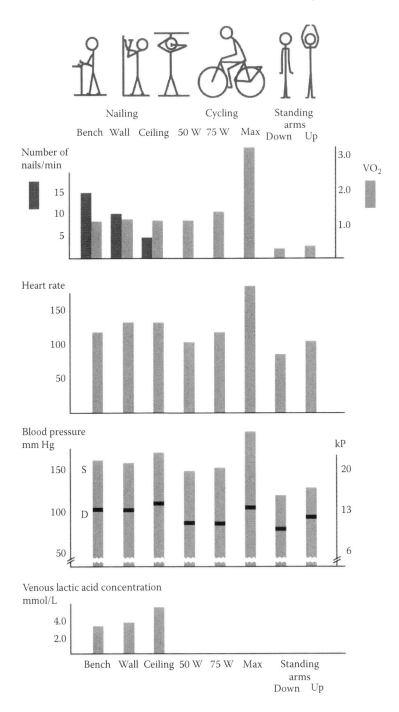

FIGURE 2.10 The reaction when hammering nails in different body postures, during cycling, and standing with the arms alongside the body or with arms lifted above the head. Despite a constant oxygen uptake (VO_2 L/min) of ~1 L/min in the different body postures when

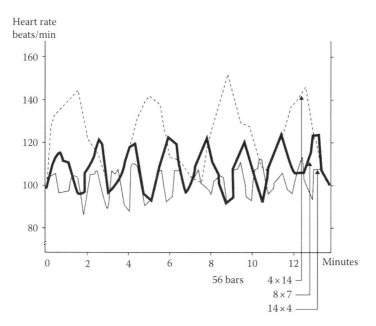

FIGURE 2.11   The heart rate of an individual loading four lots of 14 iron pieces (broken line), eight lots of seven iron pieces (thick line) and 14 lots of four iron pieces (thin line). The heart rate was appreciably lower in short work cycles compared with long cycles, despite constant production. (Adapted from Åstrand I. 1990. *Arbetsfysiologi.* 4th ed. Stockholm: Norstedts förlag (in Swedish), p. 84. With permission.)

marked drop in heart rate and lactic acid concentration if the length of the break was extended beyond a certain minimum, presupposing that the work period was short. The conclusions are that the work period must be as short as possible, so that the load is as low as possible, while the length of the break is not as critical to the outcome. The explanation of the results is presumably that, when an individual works with a high load for a short period, the oxygen supply is adequate, despite an inadequate oxygen supply during the activity. One possible explanation for this is that the myoglobin supplies the muscles with oxygen at the beginning of each work period. Myoglobin, which is to be found in all muscle cells, is closely related to haemoglobin,

FIGURE 2.10   Continued. hammering nails, production dropped from ~15–5 nails/min in the transition from nailing into a bench to nailing into the ceiling. Efficiency therefore dropped. The heart rate when cycling at 50 watts was ~100 beats/min and when nailing the ceiling ~130 beats/min despite the same oxygen uptake. Systolic blood pressure (S) was ~150 (20.0) and 170 mm Hg (22.6 kPa), respectively, and diastolic pressure (D) was ~80 (10.6), and 110 mm Hg (14.6 kPa), respectively. The venous lactic acid concentration was at its maximum when hammering nails into the ceiling. Note that diastolic blood pressure was higher when standing at rest with the arms raised above the head than when cycling at 75 W. At maximum performance on the cycle, diastolic blood pressure was lower than when nailing into the ceiling. (Adapted from Åstrand I., 1990. *Arbetsfysiologi.* 4th ed. Stockholm: Norstedts förlag (in Swedish), p. 78. With permission.)

and has the ability to bind oxygen. Over a short work period of at most 30 s, the oxygen needs are presumably covered by the amount of oxygen bound to the myoglobin. During the break the myoglobin is then loaded with new oxygen in just a few seconds. In principle, the work can continue aerobically for an indefinite time, as long as the work periods are not so long that the oxygen reserves in the myoglobin become exhausted, because then anaerobic energy metabolism takes over and lactic acid is formed.

In many forms of heavy work, where the employees themselves can regulate their work rate, putting in frequent short breaks to reduce the load works very well. In machine-controlled occupations it is not always possible for employees to influence the pattern of work and breaks, but the work can be controlled according to the principles mentioned above so as to reduce load.

It is important not to confuse the above-mentioned short breaks with other, longer breaks for recuperation, which are motivated by other reasons, both from a work environment viewpoint as well as a social viewpoint. Apart from a proper lunch break and a break for a snack both in the morning and afternoon to replenish energy stores, John needs to take breaks fairly often to drink water. He needs to drink at least half a litre per hour, and more if it is hot outside, so as not to impair his performance. If John does not drink enough, dehydration sets in; this results in a lower stroke volume at a certain work load, which is compensated for by an increased heart rate. This results in greater exertion and a reduction in performance.

2.9.1.4 Equipment

The design of equipment and tools can be very important both for the load on the individual and for productivity. In many cases, it is possible with simple means to improve the equipment so as to improve efficiency (Figure 2.12a and b). Simple measures for the bicycle messenger John are, for example, to adjust his saddle height optimally (to reduce static load on thigh muscles and improve efficiency) and to ensure that his tyres are correctly inflated. A cycle with a large number of gears means that John can optimize his efficiency, for example, when cycling up and down hills. For the slaughterhouse worker Janis (see Chapter 5), simple measures, such as keeping his cutting tools sharp, affect the load considerably.

2.9.2 ENVIRONMENTAL FACTORS

2.9.2.1 Ambient Temperature

Ambient temperature affects physical performance (see also Chapter 9). Work in a hot climate increases the cardiovascular load on the individual at a specific external exposure. Heart rate at a submaximal work load increases as a result of the greater blood circulation to the skin in order to transport heat from the working muscles to the skin for heat transfer. Prolonged heat exposure may lead to dehydration as a result of increased sweating, if the water loss is not replaced. Fluid deficit alone increases heart rate at a certain work rate, and performance is further reduced. Even with a fluid deficit corresponding to 1% of body weight, the capacity to work declines measurably. Prolonged heat exposure also impairs mental and neuromuscular function. This impairment can result in a greater risk of misjudgments and slips, thereby

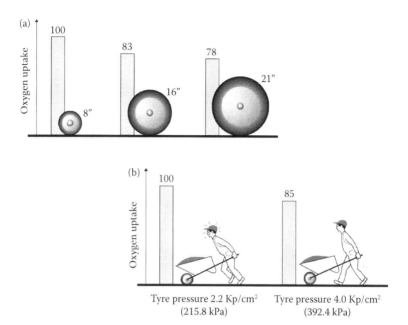

FIGURE 2.12 At a constant-sized load and transport speed oxygen uptake is lower when using wheelbarrows with large wheels compared with small ones (a) and with a high tire pressure compared with low (b). Oxygen uptake with small wheels and badly inflated tyres, respectively, is set at 100, and oxygen uptake using larger wheels and a greater tyre pressure is expressed as a percentage of the oxygen uptake measured in the worst circumstances. (Adapted from Hansson J.-E. 1970. *Ergonomi vid byggnadsarbete.* Research Report no. 8: Byggforskningen, State Council of the Building Industry (in Swedish).) Illustration: Niklas Hofvander.

increasing the risk of accidents. It is therefore important to assess heat exposure to ensure optimum conditions for health and productivity.

Heavy work in cold conditions is primarily a problem before there is time to warm up the muscles. As blood circulation and nerve conduction velocity are lower in cold tissues, the function in the musculoskeletal system is impaired, which reduces the individual's physical capacity and increases the risk of injury (see also Chapters 3 and 9).

2.9.2.2 Humidity

High humidity impedes physical capacity and increases the cardiovascular load on the individual at a specific level of external exposure, as it is more difficult to retain the heat balance when heat transfer through the evaporation of sweat decreases (see also Chapter 9).

2.9.2.3 Height Above Sea Level

Work at high altitude impedes physical capacity. Oxygen pressure in the air we inhale, which is normally at ~760 mm Hg at sea level, declines the higher we go. At 3000 m above sea level, for example, the oxygen pressure has decreased to about 500 mm Hg. The lower oxygen pressure in the air at high altitude results in a lower oxygen concentration in arterial blood, and thereby a lower arteriovenous oxygen difference. In order to compensate for the lower level of oxygen extraction, pulmonary ventilation and heart rates are higher at a certain submaximal work load in people who are not acclimatized, that is to say before the body has had time to adapt to the high altitude. Maximal aerobic capacity results from maximum cardiac output and maximum arteriovenous oxygen difference (according to Fick's principle, see Section 2.8). Maximum cardiac output is, however, the same irrespective of altitude, and as in principle all oxygen is extracted from the blood passing the working muscles at maximum work rate, VO_2 max is reduced, and thereby physical capacity, before the body has had time to acclimatize to the high altitude.

During a long stay at high altitude, that is to say acclimatization, several physiological changes occur in order to compensate for the lower oxygen pressure in the air inhaled. At a prolonged stay at high altitudes, physiological adaptations to compensate for the reduced oxygen pressure in the inspired air, that is acclimatization, takes place. Over the first few days, pulmonary ventilation continues to increase. The haemoglobin concentration gradually increases after various periods of acclimatization at high altitude, so that the oxygen content per litre of arterial blood may be the same in an acclimatized individual at high altitude as at sea level. This means that heart rate at a specific submaximal work load begins to drop, and gradually reaches the same or even a lower level compared with the rate at sea level. Other physiological changes also occur, for example, increased capillary density and increased myoglobin concentration in the muscles as well as altered enzyme activity. The initial reduction in VO_2 max is gradually recovered and is, for example, at 3000 m reduced by 5–10% in acclimatized individuals compared with ~20% for non-acclimatized individuals.

2.9.3 INDIVIDUAL FACTORS

The differences in performance capacity between different individuals are quite considerable. As much as 70% of physical capacity is thought to result from genetic factors. Some of the differences can be explained by differences between the genders and between different age groups, and also differences in body size. The major differences existing between individuals within each group involve, however, considerable overlaps between the groups. Various medical conditions, for example, heart and lung disease may of course appreciably impair capacity, but this is not dealt with in detail here.

2.9.3.1 Gender

Women's maximal aerobic capacity, VO_2 max in L/min, is after puberty on average 65–75% of that of men (Figure 2.13). The relatively large individual distribution should, however, be noted. Out of all the individuals of a certain age group, ~2.5% have 25% lower aerobic capacity and 2.5% have 25% better aerobic capacity than the average for the group. The distribution also implies that ~2.5% of all men have an aerobic capacity that is lower than that of the average woman.

The differences between men and women results, in part, from differences in body size. On average, VO_2 max increases with increased body weight raised to 2/3. When we take into account size, women's VO_2 max/kg of body weight is on average

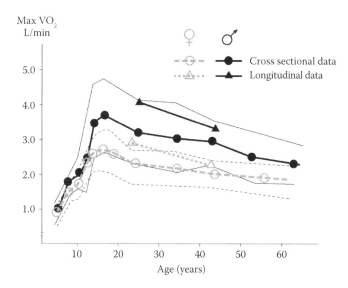

FIGURE 2.13 Average values for maximal aerobic capacity (VO$_2$ max L/min) measured in 350 normal people from the ages of 4–65 running on a treadmill or cycling on a cycle ergometer. Filled circles with solid lines represent men and empty circles with broken lines represent women. The thin solid lines and broken lines represent two standard deviations for men and women, respectively. Filled and unfilled triangles represent 31 male and 35 female students, respectively, training to be physical education teachers, who were measured during their training and again 20 years later. (Adapted from Åstrand I. et al. 1973. Reduction in maximal oxygen uptake with age. *J. Appl. Physiol.* Nov. 1, 35:649–654. Bethesda, MD: American Physiological Society. With permission.)

75–80% of that of men. This is primarily because of the greater proportion of fatty tissue, which has a low-energy metabolism, in women (~12 kg compared with 8 kg in a normal woman or man, respectively). A number of studies show that if VO$_2$ max is expressed per kilogram of fat-free body weight (what is called lean body mass), there are no differences between men and women. As oxygen uptake increases with body weight raised to 2/3, women should have a somewhat higher VO$_2$ max expressed per kilogram of fat-free body weight. The observed lower VO$_2$ max than expected in women can be explained by the lower haemoglobin concentration and thereby lower maximum arteriovenous oxygen difference.

The average lower maximal aerobic capacity in women compared with men usually means that women use a greater proportion of their maximum capacity at a specific energy metabolism. This means that a job requiring a specific energy metabolism—that is, oxygen uptake—is carried out at a higher heart rate and with greater exertion by the average woman compared to the average man.

2.9.3.2 Age

The maximal aerobic capacity increases with increased growth during childhood up until puberty, whereupon there is a gradual decline with increased age (Figure 2.13). At the age of 65, VO$_2$ max is on average ~70% of the value of that of a 25-year-old.

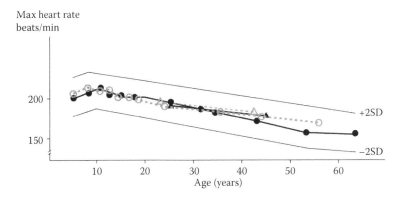

FIGURE 2.14 Average values for maximal heart rate (beats/min) measured in 350 normal people from the ages of 4–65 running on a treadmill or cycling on a cycle ergometer. Filled circles with solid lines represent men and empty circles with broken lines represent women. Filled and unfilled triangles represent 31 male and 35 female students, respectively, training to be physical education teachers, who were measured during their training and again 20 years later. The thinner solid lines represent two standard deviations (one standard deviation is ~10 beats/min for all age groups). (Adapted from Åstrand I. et al. 1973. *J. Appl. Physiol.* 35:649–654 Bethesda, Maryland: American Physiological Society. With permission.)

Therefore, VO_2 max for an average 65-year-old man is approximately the same as that for an average 25-year-old woman.

There are many reasons for this decline in aerobic capacity with increasing age. Changes in lifestyle, for example, a reduction in physical activity, are presumably a contributing factor, but physiological changes also take place with increasing age. Pulmonary ventilation becomes less effective, and the gas exchange in the lungs occurs more slowly. Additionally, maximal heart rate declines appreciably and thereby also the heart's maximum minute volume (Figure 2.14).

Apart from the cardio-respiratory changes occurring with increased age, muscle mass and maximal muscle force are reduced (see Chapter 3).

2.9.4 LIFESTYLE FACTORS

Our living habits—for example, diet, physical activity and exercise, and sleep, as well as the use of tobacco and alcohol—and our entire lifestyle affect our health and well-being as well as our physical capacity (see also Chapter 6). In this section we describe the short-term effects on physical capacity of physical exercise and the use of tobacco. As regards the significance of lifestyle for the more long-term health consequences, we refer the reader to reference literature in this field (see Further Reading).

2.9.4.1 Physical Activity and Exercise

Physical capacity is influenced by the degree of physical activity and exercise. Physical exercise can be defined as all forms of repeated physical activity that improve or maintain endurance, strength, mobility, and/or coordination. In order for the exercise to be effective, it needs to have sufficient intensity, frequency, and duration and that the time for recuperation between exercise sessions is sufficient, as physiological adaptation to greater demands occurs during the rest period. For more detailed information and advice regarding physical exercise, the reader is referred to textbooks on training physiology (e.g., Åstrand et al. 2003).

In order to achieve the best training effect as regards aerobic capacity, it is important that exercise is dynamic and involves large muscle groups, such as in walking/jogging, cycling, or swimming. The increase in maximal aerobic capacity after regular exercise results from both a central adaptation through greater stroke volume and a peripheral adaptation through greater oxygen extraction in the active muscles, that is to say the arteriovenous oxygen difference increases. The increase in stroke volume results in a lower heart rate at a particular work load and cardiac output as well as a higher maximum cardiac output (cf. Fick's principle, see Section 2.8). The higher maximal cardiac output results only from an increased maximal stroke volume, while the maximal heart rate is not affected by exercise. The increase in the arteriovenous oxygen difference is achieved through changes in the muscles involved, so that the muscles adapt to the work at a higher load. The adaptation comprises a number of effects, such as the content of enzymes in the muscle cells, sensitivity to hormones, increased capillary density in the muscles, and increased muscle mass.

2.9.4.2 Use of Tobacco

Apart from the serious health risks, smoking also results in marked acute effects on physical capacity. Smoking leads to an increase in respiratory resistance by two to three times the normal value even after one or two cigarettes. This is not noticed so much during rest, as not so much air is needed then; but during physical work, when pulmonary ventilation increases, the smoker usually feels more breathless during exertion. Smoking also affects blood circulation. Heart rate at a certain submaximal oxygen uptake has been shown to be 10–20 beats/min higher after smoking one or two cigarettes, and the difference in heart rate between smokers and non-smokers is greater the higher the work load. Both the nicotine and carbon monoxide in tobacco smoke affect blood circulation. Nicotine use results in reduced peripheral circulation through the blood vessels contracting (vasoconstriction), increased heart rate and blood pressure at a certain work load, and effects on hormone secretion. Tobacco smoke contains up to 4% carbon monoxide, the ability of which to bond with haemoglobin is ~225 times higher than that of oxygen. Even a low carbon monoxide level impairs the blood's ability to transport oxygen (through reducing the arteriovenous oxygen difference because of a lower oxygen content in arterial blood), which appreciably reduces physical capacity.

2.10 MEASURING OCCUPATIONAL ENERGY DEMANDS

In order to determine whether the energy demands of an occupation are unacceptably high, one must be able to perform a more detailed assessment than is possible with the normal values reported for various occupational groups (Table 2.3).

TABLE 2.3

Classification of Physical Work Load for Various Occupations Based on Oxygen Consumption and Corresponding Cycle Ergometer Load

Classification	Cycle Ergometer (W)	Oxygen Uptake L/min	METs	Heart Rates (Beats/min)	Occupation
Very heavy labour	>125	>1.75	>6.7	>150	Heavy manual forestry, heavy manual transport labour, firefighting with breathing apparatus
Heavy labour	100–125	1.5–1.75	5.7–6.7	130–150	Heavy construction work, agricultural labour
Moderately heavy labour	50–100	1.0–1.5	3.8–5.7	100–130	Heavy healthcare work, construction work, service, and cleaning work (hotel and restaurant)
Light labour	40–50	0.75–1.0	2.8–3.8	80–100	Household work, light factory work, light healthcare work, laboratory work, retail work
Very light labour	20–40	0.5–0.75	1.9–2.8	70–80	Office work, car driving, seated work (reading, writing)
	<20	<0.5	<1.9	<70	

Source: Data from Åstrand I. 1990. *Arbetsfysiologi.* 4th ed. Stockholm: Norstedts förlag (in Swedish); Fallentin N. 1995. *Arbejdsfysiologi*, pp. 118–135. Köpenhamn: Arbetsmiljöinstituttet (in Danish).

Note: The table also indicates the corresponding METs (for individuals weighing 75 kg), average heart rate variation and examples of occupations within the respective categories. Note that work load classifications based on oxygen consumption and corresponding heart rate values refer to an average 20- to 30-year-old, and that the considerable variation among individuals must be taken into account.

2.10.1 TECHNICAL MEASUREMENTS

2.10.1.1 Oxygen Consumption

Measuring oxygen consumption has proved to be a reliable method for assessing energy metabolism and therefore represents one acceptable way of assessing occupational demands from an energy perspective. Oxygen consumption can be determined by calculating the difference between the measured oxygen content in inhaled and exhaled air.

Portable equipment is available that can perform continuous oxygen consumption monitoring in the field and which weighs only a few kilograms (see Chapter 6, Figure 6.9). The results can be stored on a portable unit or wirelessly transferred to a computer, allowing for real-time tracking on a display screen. The energy demands of a number of different physical activities have been measured and documented [Ainsworth et al. 1993, 2000]. Johns' oxygen consumption in his job as a bicycle messenger was measured over three different working days, and his average oxygen consumption was found to be 1.54 L/min.

Although the equipment and methods have been simplified in recent years, measuring oxygen consumption in the field is not an easy method for practical uses and is more suitable for research purposes instead. The equipment is relatively expensive, and special skills are required to calibrate and operate the equipment to obtain accurate results.

2.10.1.2 Pulmonary Ventilation

For occupations that are physically not very intensive, measuring pulmonary ventilation with a flow meter is a relatively easy way of estimating oxygen consumption. Pulmonary ventilation increases linearly with increased oxygen consumption from a rest state up to a pulmonary ventilation of 40–50 L/min (Figure 2.7). This means that occupational energy demands entailing oxygen consumption below 1.75–2 L/min can be measured based on pulmonary ventilation. John's average pulmonary ventilation in his job as a bicycle messenger is ~33 L/min, which corresponds to an oxygen consumption of 1.5 L/min (Figure 2.7).

2.10.1.3 Heart Rate

Today, one can find user-friendly and relatively inexpensive equipment that can be used to continuously monitor heart rate over an entire working day with a good degree of accuracy (see Chapter 6, Figure 6.9). Assessing energy demand based on monitoring heart rate is possible because there is a linear correlation between work load and oxygen consumption (i.e., a constant efficiency factor) and between heart rate (within a certain range) and oxygen consumption. In order to perform this type of measurement, however, a so-called "biological calibration" is required. This calibration is usually performed on a cycle ergometer. The individual's own correlation between work load/oxygen consumption and heart rate is determined at least two, but preferably three, submaximal loads within the heart rate interval anticipated for the work in question (Figure 2.15). Based on the individual's (usually) linear correlation, oxygen consumption can then be estimated based on heart rate as measured at a steady state during a work task. John's average heart rate during his work as a bicycle messenger is 96 beats/min, which for him corresponds to an oxygen consumption of 1.55 L/min and a work load of just over 100 W (Figure 2.15).

An important limitation on the use of heart rates as a measure of energetic work load is, however, that the heart rate, as previously mentioned, is affected by a number of factors other than oxygen uptake alone. In heat exposure and work including static load as well as work by small muscle groups, the heart rate increases and alters the "normally" linear correlation between heart rate and oxygen uptake (Figure 2.16). At low-to-average physical load, the heart rate also increases as a result of mental stress. For a reliable assessment of energetic work loads on the basis of heart rate measurements, one should therefore carry out the biological calibration under conditions which to the greatest extent possible are similar to actual working conditions. Heart rate in itself is, however, a good measure of exertion on the individual.

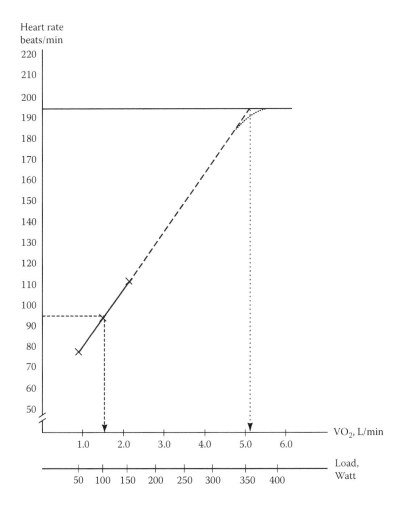

FIGURE 2.15 Biological calibration of the correlation between work load/oxygen consumption and heart rate. John's heart rate was measured at three work loads (50, 100, and 150 watts) on a cycle ergometer (× connected with a solid line). During his workday as a bicycle messenger, John's average heart rate is 96 bpm, which corresponds to an oxygen consumption of ~1.55 L/min (see ----- for the correlation between John's working heart rate and oxygen consumption). John's maximum oxygen uptake (VO_2 max) was measured at 5.1 L/min, meaning that he is using ~30% of his VO_2 max ($1.55/5.1 \times 100$) while working as a bicycle messenger. John's measured maximum heart rate was 195 beats/min and in the figure John's calibration line has been extrapolated linearly to his maximum heart rate (–––––). If it is not possible to measure a person's VO_2 max L/min, it can be predicted on the basis of the point of intersection between the person's extrapolated calibration line and maximum heart rate (see ······ for the correlation between John's maximal heart rate and VO_2 max L/min). In certain people the calibration line deviates with very heavy work (·······), that is to say the oxygen uptake increases relatively more than the heart rate, which means that VO_2 max L/min of these people will be underestimated somewhat. For instance, VO_2 max for a person whose increasing heart rate in reality deviates, according to the dotted line in the figure, will be underestimated by ~7% (predicted VO_2 max 5.1 L/min compared with real VO_2 max 5.5 L/min).

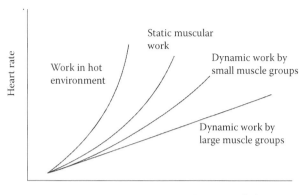

FIGURE 2.16 With increasing energy metabolism, the heart rate increases more steeply in simultaneous heat exposure and with elements of static muscular work and work by small muscle groups compared with dynamic work by large muscle groups. (Adapted from Kroemer KHE., Grandjean E., 1997. *Fitting the Task to the Human. A Textbook of Occupational Ergonomics.* 5th ed. London: Taylor & Francis, p. 116. With permission.)

2.10.2 OBSERVATION OR SELF-REPORTING

As an alternative to technical measurement, energy demands at work can be assessed on the basis of observation or self-reporting of various tasks/activities. On the basis of published reference values for energy demands for various activities [Ainsworth et al. 1993, 2000] and the total time taken for each activity, one can calculate time-weighted average values over a working day [Edholm 1966; Pernold et al. 2002]. John has kept a diary of his activities for a working week. He usually cycles for 6.5 h, at an average speed of ~18 km/h, which corresponds to ~6 METs. For the rest of the time he collects and delivers various items of mail and awaiting new tasks. Collecting and delivering items of mail takes about 60 min of each working day, and then John is often jogging both across a flat surface and a great deal up and down stairs, which on average corresponds to ~7 METs. For 30 min he is standing waiting for new tasks, and chatting to his employer and customers, ~1.8 METs [Ainsworth et al. 1993]. His time-weighted average value on a normal working day is

$$(6.5 \times 6) + (1.0 \times 7) + (0.5 \times 1.8) \text{ h} \times \text{METs}/8 \text{ h} = 5.9 \text{ METs}.$$

For John, who weighs 75 kg 5.9 METs corresponds to an oxygen uptake of 1.55 L/min $(5.9 \times 3.5 \text{ mL } O_2 \times \text{kg}^{-1} \times \text{min}^{-1} \times 75 \text{ kg} = 549 \text{ mL/min or } 1.55 \text{ L/min})$ (see Section 2.6).

2.11 DEMANDS OF WORK IN RELATION TO PHYSICAL WORK CAPACITY

The great variation in physical capacity implies that a certain task can be carried out with low exertion for certain individuals, while the same job can be very strenuous for others. Generally speaking, a specific external exposure is more strenuous for women compared with men and for older people compared with younger ones, even if the variations within the respective gender and age group are considerable. In order to be able to evaluate whether a certain work load is unacceptably high, we have to take into account the individual's maximal work capacity. The relative load expresses the relation between the demands of work and the individual's maximal work capacity, and can be expressed as

Relative load (%) = 100 × (Demands of work/individual's maximal work capacity)

2.11.1 RELATIVE OXYGEN UPTAKE

Methods for measuring work demands, expressed as demands for oxygen uptake, are described in the previous section. Individual physical capacity, expressed as maximal aerobic capacity (VO_2 max/min), can be measured directly through a so-called maximal test or indirectly on the basis of a submaximal test.

2.11.1.1 Maximal Test

In a maximal test on a cycle ergometer or treadmill, the work load is gradually increased and oxygen uptake measured at each stage. Oxygen uptake increases linearly with an increase in work load until the level is reached where oxygen uptake does not increase further despite the fact that the load increases (Figure 2.4). The highest oxygen uptake that can be measured corresponds by definition to the individual's maximal aerobic capacity, VO_2 max. John has undergone a maximal test in a work physiology laboratory. His maximal aerobic capacity was 5.1 L/min, that is to say an increase of 19 times the basal metabolic rate (19.4 METs), which is assessed as being a very high physical capacity. In the previous section, we stated that John's oxygen uptake was 1.54 L/min on average during his working day working as a bicycle messenger, which therefore corresponds to ~30% (1.54/5.1 alt. 5.9/19.4) of his maximal aerobic capacity.

John's friend Paul, who is also 26 years of age, has also had his maximal aerobic capacity tested. Paul has a sedentary job as a computer technician and on an average spends 12 h a day sitting in front of a computer screen. Paul drives to and from work and takes no physical exercise in his leisure time. His primary interest apart from computers is in films, and his leisure time is devoted mostly to watching TV and going to the cinema. His maximal aerobic capacity was 2.5 L/min, that is, half that of John, which at his age is a very low level of fitness. If Paul were to cycle at the same speed as John, that is, to say with an oxygen uptake of ~1.5 L/min, he would have to use 60% (1.5/2.5) of his maximal capacity, which would be very strenuous for him. It would be impossible for him to maintain that tempo for a whole working day. If Paul on the other hand were to cycle with the same relative load as John, his lower tempo would result in a considerably lower performance than John's.

In most physical activities, body weight in itself involves a load, for example, in walking and running. Then it is more relevant to express the maximal aerobic capacity related to kilogram of body weight; in millilitres of oxygen per kilogram of body weight and minute ($mL \times kg^{-1} \times min^{-1}$). John weighs 75 kg and his aerobic capacity related to kg body-weight is 68 $mL \times kg^{-1} \times min^{-1}$ ($5100 \; mL \times 75 \; kg^{-1} \times min^{-1}$). If John had instead weighed 100 kg, this value would have been 51 $mL \times kg^{-1} \times min^{-1}$. When John is collecting and delivering items of mail, he often jogs to and from his bicycle, which corresponds to ~7 METs, that is to say an oxygen consumption of ~24.5 $mL \times kg^{-1} \times min^{-1}$ ($7 \times 3.5 \; mL \times kg^{-1} \times min^{-1}$). This corresponds to 36% (24.5/68) of his maximal aerobic capacity in millilitres of oxygen per kilogram of body weight and minute. If John on the other hand weighed 100 kg, the same external work (exposure) would have corresponded to 48% (24.5/51).

Tests to determine maximal aerobic capacity presuppose laboratory conditions with the need for specially trained staff and specialist equipment. If we do not have the opportunity to measure the maximal aerobic capacity, we can predict it with a reasonable degree of accuracy on the basis of the individual's maximal heart rate and biological calibration line for submaximal loads (Figure 2.15). John's maximal heart rate, when his work load was increased in stages until his heart rate did not increase any further despite a higher load, was 195 beats/min. If we extend John's biological calibration line to his measured maximal heart rate, the reading will be a corresponding maximal aerobic capacity of ~5.1 L/min. For some people, however, oxygen uptake increases, relatively speaking, more than heart rate during very heavy work, which means that the maximal aerobic capacity of these individuals will be slightly underestimated (Figure 2.15).

2.11.1.2 Submaximal Test

In workplace studies, submaximal tests are usually used to predict maximal aerobic capacity. These tests are usually carried out on a cycle ergometer and do not require the maximal performance, which is why they are considerably easier to conduct. Most submaximal tests are based on the fact that there is a linear correlation between oxygen uptake and heart rate, which is common. Oxygen uptake in certain individuals increases relatively speaking more than heart rate during very heavy work, which is why the maximal aerobic capacity in such cases will be underestimated. Submaximal tests also presuppose a constant mechanical efficiency similar to all individuals. Certain people manage a certain work load with a lower energy metabolism than others, and people who are very much overweight have a higher oxygen uptake at a certain load. The differences in efficiency in cycling are relatively small compared with running, where the technique and the body weight have a greater significance for energy metabolism. Furthermore, it is presupposed that the maximal heart rate depends only on age, despite the fact that the distribution within a certain age group is relatively large (Figure 2.14).

In order to calculate an individual's maximal aerobic capacity, one can measure their heart rate (after 3–4 min, that is, to say when the steady state has been achieved, see Figure 2.5) at two or three submaximal rates at least. If a straight line is adjusted to the results describing the relationship between heart rate and work load, this line can then be extended to the value for a predicted maximal heart rate based on age. The individual's maximal aerobic capacity can then be assessed in a corresponding way to that shown in Figure 2.15, with the difference that the results in the figure are based on a measured, and not predicted, maximal heart rate. The decline in maximal heart rate with increasing age is evident from data in Figure 2.14 described with the formula

$$\text{Maximal heart rate} = 220 - \text{age [Åstrand et al. 2003]}$$

Another study shows a somewhat smaller age-related decrease in the maximal heart rate, described with the formula

$$\text{Maximal heart rate} = 210 - (0.662 \times \text{age}) \text{[Bruce et al. 1973]}$$

On the basis of the formulae given above, John's predicted maximal heart rate is 194 and 193 beats/min, respectively. His predicted maximal aerobic capacity on the basis of a submaximal test is ~5.0 L/min (compare with the process in Figure 2.15). In John's case, the difference between the predicted and measured maximal heart rate is small (193 or 194 compared with 195 beats/min) and his predicted VO_2 max does not differ very much from the measured value (5.0 compared with 5.1 L/min). The difference between the predicted and measured maximal aerobic capacity is often greater than it is in John's case. On the basis of the submaximal test of VO_2 max above and his predicted oxygen uptake, based on average heart rate during work as a bicycle messenger (see Section 2.10.1 about heart rate), John's work as a bicycle messenger is predicted as corresponding to ~31% (1.55/5.0) of his VO_2 max.

Another common method for predicting the maximal aerobic capacity on the basis of the cycle ergometer test is based on the Åstrand and Åstrand nomogram. Here the heart rate is only registered at one submaximal load (~140 beats/min for subjects <50 years of age and ~120 for older subjects) and the maximal aerobic capacity can then be calculated on the basis of heart rate at the chosen work load. The value obtained then has to be corrected for age [Åstrand et al. 2003]. The nomogram is based on empirical research into a very large number of people, and the correlation between heart rate at a submaximal work load and maximal aerobic capacity is therefore fairly good at the group level. When John carried out a cycle ergometer test at the load of 200 watts, his heart rate at steady state was on average 132 beats/min. After correction for age (reading value × age factor) John's VO_2 max is estimated at 5.2 L/min.

On the basis of Åstrand's submaximal test of John's VO_2 max and his predicted oxygen uptake, based on average heart rate when working as a bicycle messenger (see Section 2.10.1 about heart rate), John's work as a cycle messenger is estimated to correspond to ~30% (1.55/5.2) of his VO_2 max.

At the individual level the results of submaximal tests must be evaluated with care, as the error is relatively great, which results in the predicted value of VO_2 max not always corresponding to the real (measured) value. This error results primarily from the great distribution in maximal heart rate even within a particular age group (see Figure 2.14). The error means that 95% of the individuals in a certain age group have a maximal heart rate which lies within the limits of ±20 beats/min from the average value. This means, for example, that 2.5% of a group of 26-year-olds have a maximum heart rate higher than 213 beats/min (193 + 20) and 2.5% have a maximum heart rate lower than 173. Individuals with a higher maximal heart rate than the average will be underestimated and vice versa (see Figure 2.15 and compare the predicted VO_2 max at different maximal heart rates). This source of error can be corrected if we know the individual's real maximal heart rate and use that in the prediction instead of a group average value. Another source of error is the variation in efficiency. This error means that 95% of the individuals in a group have an oxygen uptake that lies within the limits of ±12% of the average value at the respective work load. This results in 2.5% of the group of people tested at, for example, 150 W, which on average requires an oxygen uptake of 2.1 L/min, have an oxygen uptake higher than 2.35 L/min [2.1 + (0.12 × 2.1) L/min] and 2.5% have an oxygen uptake lower than 1.85 L/min [2.1 − (0.12 × 2.1) L/min]. An individual who has an oxygen uptake of 1.8 L/min at 150 W has a very high efficiency, and the individual's predicted VO_2 max will be overestimated, while a person with low efficiency will be underestimated. A further source of error is that heart rate in some individuals does not increase linearly with a rise in oxygen uptake all the way up to the maximal. For individuals whose heart rate increases relatively speaking less than their oxygen uptake during heavy work, the predicted VO_2 max will be underestimated somewhat (Figure 2.15).

The sources of error listed above mean that the test result is incorrect for many people. The size of the error in Åstrand's submaximal cycle ergometer test means that 95% of all those individuals tested lie within the limits of ±30% of the actual value. This means, for example, that for 2.5% of the group of individuals whose VO_2 max according to Åstrand's test is predicted to be 5.2 L/min, the real value lies below 3.6 L/min [5.2 − (0.3 × 5.2) L/min] and for 2.5% the real value lies above 6.8 L/min [5.2 − (0.3 × 5.2) L/min] (even if such a high maximal aerobic capacity is very unusual). If the test is carried out according to the standardization requirements, the results have good reproducibility; that is, the degree of deterioration or improvement, respectively, of an individual's VO_2 max can be followed with high precision, even if the absolute value is subject to relatively high uncertainty.

2.11.2 RELATIVE HEART RATE INCREASE

The relative load can also be described on the basis of heart rate. In order to compensate for the age-related reduction in maximal heart rate, the relative load is best expressed as a relative heart rate increase, that is, the difference between the heart rate increase at work divided by the maximal heart rate increase (the heart rate reserve).

Relative heart rate increase (%) = 100 × (heart rate at work

− heart rate at rest)/(maximal heart rate

− heart rate at rest)

John's heart rate was on average 96 beats/min when working as a bicycle messenger; his maximal heart rate was 195 and rest value 55 beats/min, which means a relative heart rate increase of 29% [(96−55)/(195−55)]. If we assume that Paul's heart rate at the same cycling speed was ~140 beats/min, and that he has the same maximal heart rate as John but a rest pulse of 60 beats/min, then his relative heart rate increase at the same work load is 56% [(135−60)/(195−60)].

If John were 60 years old instead (a predicted maximal heart rate of 170 beats/min), a heart rate of 96 beats/min would be more strenuous and correspond to a relative heart rate increase of ~36%.

2.12 EFFECTS OF HIGH-ENERGETIC LOAD

2.12.1 SHORT-TERM EFFECTS

Both physical and mental fatigue are common short-term effects of working at high energy metabolism. Fatigue is defined as a state of disturbed equilibrium (homoeostasis) and can result in both subjective and objective symptoms as well as impaired ability to work. Physical exhaustion can be local and/or central, and sometimes aches/pains occur in combination with a general feeling of fatigue. Both aches/pains and fatigue should be interpreted as a warning signal to prevent overload, and work should be interrupted or the work load reduced to achieve recuperation. Fatigue can result in impaired sensory and motor control as well as impaired perception and cognitive functions, which can lead to poor coordination and increase the risk of overload injuries in, heavy manual work, for example.

2.12.2 LONG-TERM EFFECTS

High energy metabolism implies a physically active life, which often has many positive health effects such as a reduced risk of cardiovascular disease, obesity, diabetes, osteoporosis, and certain forms of cancer, as well as an improved immune defense system. Physical inactivity is more widespread in today's society and a greater public health problem (see Chapter 6). In most occupations with a high energy metabolism, the occurrence of heavy manual handling is common, for example, within the nursing and care professions (see Chapter 3), which is why much of what is called heavy work has an excess risk of such things as overexertion accidents and disorders of the musculoskeletal system. Many people are unable to work until they are 65 years of age, and have to take early retirement. Today there is no scientific evidence that heavy physical work generally has a positive effect on physical capacity. Some studies indicate that heavy work can even accelerate the age-related decline in physical capacity. Presumably, in most occupations the load is not optimal as regards level, duration, frequency, and, perhaps above all, recuperation to provide a central and peripheral adaptation, and therefore a training effect.

As regards John the bicycle messenger, he has chosen his job among other things to improve his physical capacity before undertaking an adventurous journey. John has the advantage that he himself can determine his work load and he has therefore been able to increase the tempo successively, whereupon his fitness has improved. When he started out as a bicycle messenger, his maximal aerobic capacity was ~4.3 L/min, and after two months had increased by almost 20% to 5.1 L/min. John organized his work such that, when road and traffic conditions allowed, he cycled at an intensity corresponding to 70–80% of its maximum ability for between half an hour and an hour, and subsequently at a slower pace for the same length of time to recuperate, and in this way he could gradually improve his maximal capacity. There are, however, few occupations where the work can provide such good preconditions for exercise. One negative effect which, however, John experiences as a result of his work is that he is often so tired after work that he cannot manage a social life. As John himself can affect his work load, he would be able to reduce the physical strain by taking it a little easier. Admittedly he would then earn a little less money, but he would have the energy left after work to cope with a social life. Many people with heavy physical occupations, for example, within the nursing and care professions, cannot affect their work load, however, which means that many people have difficulty in coping with their work when they become older and their physical capacity declines.

2.13   RECOMMENDED EXPOSURE LIMITS

The International Labour Organization (ILO) has suggested 33% of VO_2 max as the highest acceptable average load during an 8-h working day. If this cannot be achieved through organizational and/or technical measures, the ILO recommends a shorter working period [Bonjer 1971].

The proposal for an acceptable load for work with high energy metabolism is based on the fact that it should be possible to maintain the work load over an 8-h shift without the physiological balance being disturbed (maintaining homoeostasis); that is, without an accumulation of lactic acid in the blood and without an increased heart rate (when maintaining the work load) [Jørgensen 1985]. The starting point is that the work should not cause such fatigue that the individual would not have the strength for active and meaningful leisure time. The determination of the exposure limit is based on the demands of the work for oxygen uptake in relation to the individual's maximal aerobic capacity, that is, relative load. In practical situations, when there is no opportunity to measure oxygen demands in the work and the individual's maximal aerobic capacity, it is possible to assess the relative load on the basis of measuring heart rate at work, the individual's biological calibration line, and submaximal test for calculating the individual's VO_2 max (see Section 2.11.1.2). If the individual's maximal heart rate has not been measured but predicted on the basis of age, the uncertainty in these assessments is so great that the result is best suited for an evaluation at group level, while the results for specific individuals have to be evaluated with care.

In continuous dynamic work by large muscle groups, the accumulation of lactic acid does not occur if the work rate does not exceed ~40% of the individual's maximal capacity (VO_2 max) (Figure 2.6). It has also been shown that in dynamic work by major muscle groups, the individual makes use of 35–40% of VO_2 max if their work rate can be determined freely. If the work on the other hand includes manual handling, which most occupations with high energy metabolism do, the elements of static muscle work and work by minor muscle groups result in a lower efficiency. An individual's maximal aerobic capacity may, for example, drop by 20–40% in lifting work compared with dynamic cycle work. In lifting work, lactic acid accumulates at a lower oxygen uptake compared with dynamic work by large muscle groups. In order to take into account occupations including manual handling, 30–35% of VO_2 max, measured during dynamic muscle work on, for example, a cycle ergometer, has been proposed as a reasonable exposure limit for an 8-h working day [Jørgensen 1985]. In order to include most employees, the energy demands of work should not exceed an oxygen uptake for men of <40 years of age: 0.7 L/min, men over 40 years of age: 0.6 L/min, women under 40 years of age: 0.6 L/min, women over 40 years of age: 0.5 L/min [Jørgensen 1985]. It should be noted, however, that these values do not exclude local fatigue as a result of heavy manual handling, which is why any risks of accidents to and disorders in the musculoskeletal system have to be assessed separately.

John's relative load (internal exposure in relation to the individual's aerobic capacity) when working as a bicycle messenger corresponded to ~30% of his VO_2 max. If, for example, instead of being 26 years of age he had been 60 years of age (predicted maximal heart rate of ~170 according to Bruce; see Section 2.11.1 on submaximal tests), his VO_2 max would have been ~4.15 L/min, instead of 5.1 L/min (see Figure 2.15). The same external work (*external exposure*), corresponding to an oxygen uptake of 1.54 L/min would then have required 37% of his VO_2 max, and would therefore have exceeded the recommended exposure limit.

2.14   MEASURES IN WORK REQUIRING HIGH ENERGY METABOLISM

In jobs with high-energetic load, it is important that the individuals themselves are able to determine their work pace and control both their work load and breaks. It is particularly important for older workers in order to be able to continue to work up to the normal pensionable age. In very heavy work, short breaks can considerably reduce the exertion while maintaining production. In addition, there should be a review of the tasks to see whether they can be broadened to include physically less demanding operations, for example, reading and writing, so that the phases of heavy work can be alternated with tasks allowing for physical recuperation. In many work situations, energetic load can be reduced through mechanization, improved equipment and technical aids. In occupations where it is impossible to avoid elements requiring extremely high physical demands, for example, for fire-fighters in rescue work and the like, physical training must form part of the work so that a high physical capacity can be retained.

2.15   WHAT DOES THE LAW SAY ABOUT WORKING WITH A HIGH ENERGY METABOLISM?

Within the EU there are no explicit work environment rules regarding work demanding high energy metabolism. There is, however, a framework directive which in general regulates the employer's responsibility to ensure that employees can carry out their work without risk to their health [Directive 89/391/EEC]. The practical consequences of this mean that employees with jobs demanding a high energy metabolism must be given clear freedom of action as regards breaks and pauses, consumption of food and drink, and so on. Additionally, really heavy work should be alternated with lighter work, for example, administrative tasks as described in the previous section.

Another piece of European legislation which has a certain bearing on this type of work is a directive regulating the use of personal protective equipment [Directive 89/656/EEC]. In it, there are demands that protective recruitment should be designed ergonomically and adapted to the employee. It is important to reduce the physical load in occupations demanding higher energy metabolism in which personal protective equipment must be worn at the same time.

2.16 SUMMARY

This chapter deals with heavy physical work characterized by high energy metabolism and which makes great demands on the body's ability to take up oxygen. Almost 1/5 of the working population of Sweden and probably most other postindustrial countries are considered to have an occupation with high energy metabolism. Heavy physical work results in load on the respiratory and circulatory organs, and the relative load on the individual depends on how great a proportion of the individual's maximal capacity is used in the work. With a high relative load on the individual, lactic acid is accumulated during the working day, and fatigue occurs. The load on the individual in a specific amount of work performed depends on factors in the work—for example, how different operations are carried out, how the work is organized, and what equipment is used. The load is also affected by factors in the environment, for example, how hot and humid it is. Additionally, there are individual characteristics—for example, age—and lifestyle factors—for example, physical activity and training—which are of great significance for individual load at a specific exposure.

To avoid developing such fatigue that the individual has no energy to pursue active and meaningful leisure time, it is recommended that the average load during an 8-h working day should not exceed 33% of the individual's maximal aerobic capacity.

REFERENCES

Ainsworth BE., Haskell WL., Leon AS. et al. 1993. Compendium of physical activities: Classification of energy costs of human activities. *Med. Sci. Sports Exerc.* 25(1):71–80.

Ainsworth BE., Haskell WL., Whitt MC. et al. 2000. Compendium of physical activities: An update of activity codes and MET intensities. *Med. Sci. Sports Exerc.* 32(9):498–516.

Åstrand I. 1990. *Arbetsfysiologi.* 4th ed. Stockholm: Norstedts förlag (in Swedish).

Åstrand I., Åstrand PO., Hallbäck I., Kilbom Å. 1973. Reduction in maximal oxygen uptake with age. *J. Appl. Physiol.* 35:649–654.

Åstrand PO., Rodahl K. 1986. *Textbook of Work Physiology.* 3rd ed. New York, NY: McGraw-Hill Book Co.

Åstrand PO., Rodahl K., Dahl HA., Stromme SB. 2003. *Textbook of Work Physiology. Physiological Bases of Exercise.* 4th ed. Champaign, IL: Human Kinetics.

Bernmark E., Wiktorin C., Svartengren M. et al. 2006. Bicycle messengers: Energy expenditure and exposure to air pollution. *Ergonomics.* 49(14):1486–1495.

Bonjer FH. 1971. Energy expenditure. In L. Parmeggiani (ed), *Encyclopedia of Occupational Health and Safety*, pp. 458–460. Geneva: International Labour Organization.

Bruce R., Fisher LD., Cooper MN. et al. 1973. Separation of effects of cardiovascular disease and age on ventricular function with maximal exercise. *Am. J. Cardiol.* 34:546–550.

Directive 89/391/EEC—On the introduction of measures to encourage improvements in the safety and health of workers at work.

Directive 89/656/EEC—On the minimum health and safety requirements for the use by workers of personal protective equipment at the workplace.

Edholm OG. 1966. The assessment of habitual activity. In K. Evang and K. Lange Andersen (eds), *Physical Activity in Health and Disease*, pp. 187–197. Oslo: Universitetsforlaget.

Ekblom B., Hermansen L., Saltin B. 1967. *Hastighetsåkning på skridsko*. Idrottsfysiologi, Rapport no 5. Stockholm: Framtiden (in Swedish).

Fallentin N. 1995. Tungt fysisk arbejde. In G. Sjögaard (ed), *Arbejdsfysiologi*, pp. 118–135. Köpenhamn: Arbetsmiljöinstituttet (in Danish).

Hansson J.-E. 1970. *Ergonomi vid byggnadsarbete*. Research Report no. 8: Byggforskningen, State Council of the Building Industry (in Swedish).

Jones DA., Round JM. 1990. *Skeletal Muscle in Health and Disease*. Manchester and New York, NY: Manchester University Press.

Jørgensen K. 1985. Permissible loads based on energy expenditure measurements. *Ergonomics* 28(1):365–369.

Kroemer KHE., Grandjean E. 1997. *Fitting the Task to the Human. A Textbook of Occupational Ergonomics*. 5th ed. London: Taylor & Francis.

Pernold G., Wigaeus Tornqvist E., Wiktorin C. et al. 2002. Validity of occupational energy expenditure by interview. *AIHA J.* 63:29–33.

SWEA. 2009. *The Work Environment 2009*. Swedish Work Environment Authority and Statistics Sweden. Work Environment Statistics, Report 2010:3 (in Swedish with English summary).

Wigaeus Tornqvist E., Kilbom Å., Vingård E. et al. 2001. The influence on seeking care because of neck and shoulder disorders from work-related exposures. *Epidemiology* 12(5):537–545.

FURTHER READING

Nordic Council of Ministers. 2004. *Nordic Nutrition. Recommendations 2004. Integrating Nutrition and Physical Activity*. 4th ed. Copenhagen: Nordic Council of Ministers.

3 Work Requiring Considerable Muscle Force

Katarina Kjellberg

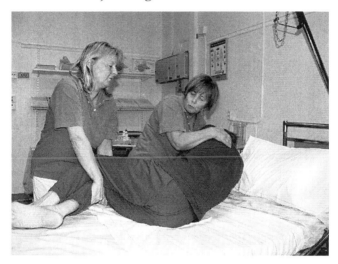

Photo: Joakim Romson

CONTENTS

Karen is 40 years old and works part-time at an orthopedic clinic in an emergency hospital. She has been working as an assistant nurse for 20 years, the last three years in the same ward. In the ward, patients who have fractures and who have undergone various types of orthopaedic operations are cared for. These patients often need a great deal of assistance during transfers, as many of them have recently had operations or are elderly. Karen assists patients during transfers about 10–15 times each day. Her work otherwise consists primarily of changing dressings, taking samples, patient bed care, taking round and collecting in meal trays, moving beds with patients to and from operations, and unpacking material in the stores. Karen enjoys her work but feels that patient transfers are heavy for her back. The ward is not big enough for so many patients requiring care. The premises are cramped and access to technical aids is poor. She also feels that her knowledge of transfer technique and the use of technical aids are inadequate. She attended an introductory course in transfer techniques when she started care work, and has since gone on short half-day courses on three occasions, most recently 5 years ago. Karen has pain in her lower back from time to time, particularly when she is assisting patients during transfers, during patient care in beds and when she is moving patients' beds. She has had short periods of absence due to illness several times a year as a result of these problems.

3.1   FOCUS OF THE CHAPTER AND DELIMITATIONS VIS-À-VIS OTHER CHAPTERS

This chapter deals with work requiring considerable muscle force—that is, a large proportion of the muscle's maximal contractile capacity. Work of this kind occurs when the employee has to overcome considerable external forces, for example, the gravitational force of a load which is being lifted, or resistance from a patient bed which has to be set in motion. This chapter describes the loads arising on the musculoskeletal system in work requiring considerable muscle force, and physiological responses to this load. High loads are often limited to the lumbar spine, the shoulder, the forearm or knee joint, for example, which is why they are called *local load*. Prolonged periods of heavy muscle work also involve demands being made on the body's ability to metabolize energy, and thereby on oxygen uptake and blood supply. This energetic load is known as *whole-body load*. How energy metabolism adjusts to increased energy demands is not dealt with in this chapter, but is described in Chapter 2.

This chapter answers questions such as:

- What factors influence the load on muscles and joints in manual handling of burdens and patient transfers?
- Is working technique of any significance?
- How should heavy muscle work be designed for an adaptation of muscle strength to occur to meet the demands at work?
- What happens in muscles and other parts of the body during prolonged heavy muscle work?
- Why does pain occur in the lower back when handling loads?
- What can Karen or her employers do to help her avoid problems in muscles and joints?
- What can Karen do to keep on with this work right up until she retires?

3.2   WHAT CHARACTERIZES WORK REQUIRING CONSIDERABLE MUSCLE FORCE?

In working life, considerable local loads arise primarily when the employee is performing heavy manual handling. Manual handling is usually defined as transferring loads where the employee, using muscle force, lifts, lowers, pushes, pulls, carries, holds, or supports an object or living being [SWEA 1998; EUR-Lex 1990]. Another common example of work requiring great muscle force is work using hand-held machines and tools.

These types of work tasks are often *dynamic* in character, that is, the muscles change their length and force when they contract. Often each task lasts only for a relatively short period, for example, during a lift, but recurs on repeated occasions during the working day. The muscle can then make use of almost 100% of its strength (*100% MVC*—see Fact Box 3.1), and relax shortly thereafter. Repeated work operations requiring somewhat less muscle force over a longer period without time for recovery are also counted as work requiring considerable muscle force, for example, lifting goods in a warehouse for a large part of the working day. Carrying loads and forceful grips when working with hand-held machines and tools may also require a great deal of muscle force. In this case it is a question of uninterrupted work with few breaks for rest over a longer period. All three types of work requiring considerable muscle force—that is, occasional peak loads for short periods, repeated loads for a longer period, and an uninterrupted load with few breaks for rest over a longer

FACT BOX 3.1

MVC, Maximal Voluntary Contraction: MVC is a measure of the individual's maximal contractile ability, or maximal strength, in a particular muscle or muscle group, or in a specific movement of the joint. MVC is measured in terms of force (N) or torque (Nm).

%MVC, percentage of Maximal Voluntary Contraction: The development of force in the muscle or muscle group, or in a specific movement of the joint, relative to the individual's maximal capacity. % MVC therefore expresses how much the individuals exert themselves in relation to their muscle strength.

period—may lead to injury as a result of overloading the musculoskeletal system (see Section 3.9.1).

Karen's work with patient transfers involves major muscle groups, particularly in the trunk and legs, and also in the arms and shoulders. Those muscles that effect the transfer itself are naturally active, but so too are muscles used to ensure both that the body is stabilized and does not yield to countervailing forces arising during the transfer, and that the legs are firmly on the ground. Joints, tendons, and ligaments are also subjected to considerable strain. Patient transfers make special demands on staff, as patients are living individuals who have to be managed. Here it is not usually a question of lifting an individual's entire body weight from the ground, but of supporting a patient who bears some of the weight on the legs, for example, when transferring between a bed and a wheelchair, or of pulling the patient higher up in the bed. Patient transfers are often tricky to accomplish from a motor point of view, and make considerable demands on staff regarding the coordination of movements and force development, as well as their ability to cooperate with colleagues and patients. Patients can be recalcitrant, and the staff have to be constantly prepared for unexpected events, such as a patient suddenly resisting or fainting. Often, care staff work in confined spaces, which may impede safe work postures and safe working technique, therefore increasing the demands on muscle force. Entirely manual lifts are necessary particularly in acute situations—when a patient unexpectedly collapses, for example—and among rescue and ambulance personnel.

Forceful grips in work with hand-held tools, for example, among plumbers and carpenters, put strain especially on the muscles of the hand and forearm.

3.3   PREVALENCE OF HEAVY MUSCLE WORK IN WORKING LIFE

Despite the fact that a great deal of the heavy work in, for example, industry has been mechanized, automated, and replaced by modern technology, heavy manual handling and other work requiring considerable muscle forces is still common in working life. These tasks will presumably never be completely replaced by, for example, mechanical aids. This applies also to work within the nursing and care professions, where manual transfer of people will never be entirely avoidable.

Heavy manual handling occurs in many commonly occurring occupations and is more frequent among men than among women. According to the 2010 European

Working Survey, 42% of working men and 24% of working women in European Union (EU) countries reported that their work involved carrying or moving heavy loads at least a quarter of the time [Eurofound 2010]. In contrast, 13% of women, but only 5% of men, reported that they lifted or moved people at least a quarter of the time in their work. Manual handling of people is thus more frequent among women than among men.

In Sweden, the proportion of the working population lifting heavy loads daily in 2009 was 10% of women and 18% of men [SWEA 2010a]. *Heavy lifting* is defined as lifting at least 15 kg several times a day. It is worth noting that the same load, for example, 15 kg, requires a greater proportion of the maximal strength for an average woman than for an average man. It is, therefore, more difficult for a woman to lift this weight. Heavy lifts are common in several female-dominated occupations, such as assistant nurses and nurses' aides within hospitals, among care assistants and personal assistants in the care sector, and among preschool staff. For both women and men, heavy lifts are common in warehouse work. Examples of other occupations where heavy lifting is common for men are agricultural work, construction work, carpentry, and work in the food industry.

No substantial change is taking place with regard to the prevalence of heavy manual handling. In Sweden, between 1995 and 2009, the proportion of women working with heavy lifts decreased from 16% to 10%, and the proportion of men from 22% to 18% [Statistics Sweden 2011]. The European figures given above for carrying heavy loads remained unchanged since 2000 [Eurofound 2010].

3.4 THE STRUCTURE AND FUNCTION OF THE MUSCULOSKELETAL SYSTEM

The musculoskeletal system forms the basis of a person's ability to perform movements and develop force. It consists of the *skeletal system* and the *muscular system*. These two systems are linked both structurally and functionally. The skeletal system consists of bones and joints. The muscular system consists of skeletal muscles and tendons. A skeletal muscle, consisting of groups of muscle bundles, runs together into tendons. The tendons penetrate the bone tissue, which is how they provide the muscle with a steady anchoring point in the skeleton. Functionally, the skeletal muscles together with the skeletal system produce motion of the body parts and move the body in space. They also help to support the body, that is, they keep the body in an upright position and stabilize it during motion or in a particular position without motion.

A muscle acts across one or several joints through being attached to different parts of the skeleton. A movement occurs in a joint by means of one or more muscles contracting. The tensile force generated by the muscle is then transmitted to the skeleton through tendons. Depending on where the muscle attaches in relation to a joint, it will either produce a bending or extending of the joint. Muscles that have a bending function, and others that have an extending function, operate across every joint. Muscles that have the same effect across a joint are called *agonists*, and those with an opposing effect are called *antagonists*. These muscles work together in complicated patterns so that appropriate movements can be carried out. Antagonists may also contract simultaneously so that no movement occurs in a joint, so-called *co-contraction*. This is a way to stabilize a joint.

Apart from the skeleton, muscles and tendons, the musculoskeletal system also consists of ligaments, joint capsules, cartilage, and nerves that innervate the muscles, so-called motor neurons.

The musculoskeletal system is, in turn, a tool in a larger system for controlling movements: *the motor system*. Also included in the motor system is the *central nervous system* (CNS). The CNS sends out signals to the muscles based on previous motor experiences and on information that the CNS acquires from sensory organs in muscles, joints, and skin, for example.

The structure and function of the muscles are described in Chapter 6, Section 6.4. The interplay between the nervous system and muscles is described in Chapter 6, Sections 6.4.4 through 6.4.6 and also in Chapter 5, Section 5.4, where the role of the CNS is explained. A basic description of the structure and function of the skeleton and connective tissue may be found in textbooks on anatomy and physiology. An important difference between muscles and connective tissue is that the former contain contractile components, which means that the muscles can contract and produce force, so-called *active muscle force*. The connective tissue also plays an important role in joint motion by transmitting and resisting tensile forces and in a stretched condition exerts force so as to return to their original length, so-called *passive forces*. There is also connective tissue in the muscle. These passive structures therefore contribute to motion and stabilization of the body parts.

3.5   LOAD ON THE MUSCULOSKELETAL SYSTEM IN HEAVY MUSCLE WORK

In manual handling work, the muscles develop force to overcome or resist external forces. In order to carry out a lift, the gravitational force of the object has to be overcome. The gravitational force of the body parts—that is, the trunk, head, and arms—also has to be overcome. In order to be able, for example, to push a hospital bed or a wheeled stretcher when moving patients, the friction force between the wheels and the floor has to be overcome. These *external forces* (*external exposure* in Figure 1.1; see also Chapter 1, Section 1.2) give rise to forces acting on and within the various structures of the musculoskeletal system (*internal exposure* in Figure 1.1; see also Chapter 1, Section 1.2). Forces acting on and within the musculoskeletal system are usually called *mechanical load*.

3.5.1   THE RELATIONSHIP BETWEEN FORCE AND MOTION

Biomechanics may be defined as the application of the principles of mechanics to the study of biological systems. The laws of mechanics are used to describe human movements and the forces acting on various body parts during movement. They are also used to describe how external forces give rise to load on the various structures of the musculoskeletal system. In this section, the basic principles of biomechanics will be explained very briefly, and are limited to those principles we need to know in order to understand the relationship between external forces and loads on the musculoskeletal system. In order to acquire a more fundamental understanding of the principles, the reader is referred to textbooks in biomechanics (see Further Reading).

According to the laws of motion, a force is required to start, stop, or change the direction and velocity of the motion. This force is proportional to the magnitude of the *acceleration* (i.e., the increase in velocity) or deceleration (i.e., the decrease in velocity). The greater the accelerations or decelerations of the motion, the greater the forces will be. This is the reason why measuring acceleration is an important component in biomechanical studies in order to calculate forces.

If an object does not move, or is moving with a constant velocity (i.e., to say acceleration is 0), then the sum of all forces affecting the object is equal to 0. This also means that, for the object to be able to be at rest or in *dynamic equilibrium*—that is, to be able to move at a constant velocity—any force must be resisted by an equal and opposite force. As an example, the *gravitational force*, that is, the force by which the Earth's force of attraction subjects all objects on its surface, is resisted by an equally large reaction force for the force of gravity not to pull the object towards the interior of the Earth.

In biomechanics we speak of *external forces* generated outside the human body and *internal forces* generated within the body. The external forces are composed of the gravitational forces acting on the objects we are working with, and the gravitational forces on parts of our own body, as well as forces from, or applied to, external objects, for example, when a person is pushing an object. The internal forces are the active muscle forces as well as passive tensions in, for example, the tendons, ligaments, fascia, and joint capsules. These internal forces may in turn cause so-called compression and shear forces acting on the joints. A *compressive force* acts perpendicular to the joint surface, pressing the joint surfaces against each other. One example is the compressive force on the disc between the fifth lumbar vertebra and the upper part of the sacrum (the L5/S1 disc), which is often calculated in assessments of load on the lumbar spine. A *shear force* acts parallel to a joint surface, that is to say, perpendicular to the compressive force, and tends to cause a joint surface to slide in relation to the other joint surface.

A *lever* is an object that can rotate about an axis. According to the lever principle, a force that acts on a lever has a pivoting or rotating effect on that lever. This applies when the force does not act through the axis of rotation of the lever. The rotating ability of the force is called the force's *moment* (or *moment of force, torque*). The effect of a force on a lever depends, on the one hand, on the magnitude of the force and, on the other, on the length of the moment arm and can be calculated as:

$$\text{Torque} = \text{Force} \times \text{Moment arm},$$

where the *moment arm* is defined as the perpendicular distance from the action line of the force to the axis of rotation. The rotating ability of the force, the moment, is applied in a specific direction. In calculations, we consider the moments being directed clockwise or anticlockwise. One precondition for an object that can rotate being in balance is that the moments acting clockwise are equal to the moments acting counterclockwise (*torque equilibrium*).

The ability of the muscle to produce motion thus depends not merely on the contractive force of the muscle, as the moment arm of the muscle has an equal significance.

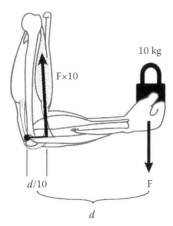

FIGURE 3.1 An example of unfavourable leverage in the muscles. When someone is holding a weight of 10 kg in their hand with a 90° angle at the elbow, the weight will cause an
external torque on the elbow joint which will extend the elbow joint. In order to maintain this
joint position, the elbow flexors have to produce an equally large internal torque. As the
weight's moment arm (d) is 10 times longer than the moment arm of the elbow flexors ($d/10$),
the elbow flexors have to develop a force that is 10 times greater than the gravitational force
of the weight. The muscles therefore have to develop a force of ~1000 N (F × 10), which may
be compared with the gravitational force of the weight of ~100 N (F). In the example, the fact
that the weight of the forearm also contributes to the external torque is not taken into account.

When the muscle pulls at a muscle attachment, the bone acts as a lever and a movement occurs in the joint. For the muscle, the moment arm comprises the right-angled
distance from the direction of pull of the muscle to the axis of rotation (Figure 3.1).

In biomechanics, we also talk about *external torque*, caused by external forces,
and *internal torque*, caused by internal forces. In order for the position of a joint to
be maintained, or a body posture to be sustained, the external and internal torques
therefore have to be equal—that is, balanced (Figure 3.1). In order to carry out a
movement, the muscles have to achieve an internal torque that exceeds the external
torque caused by gravitational forces and any other external forces. As the muscles
usually have much shorter moment arms than the external forces, the internal forces
may become very great, even with small external loads.

In the two calculation examples (Examples 3.1 and 3.2) that follow, biomechanical
calculations are made for load on the lumbar spine when a man is standing holding a
weight in front of him (Figure 3.2). The gravitational force of the load and his own body
weight cause an external torque which wants to bend his upper body forward. In order
to counteract this, the back muscles have to produce an internal torque in the opposite
direction, which is of equal magnitude to the one that wants to bend his upper body
forward. If the load is to be moved—that is, lifted—the internal torque has to exceed
the external torque. As the moment arms of the muscles are short in comparison with
those of the external forces, great muscle force has to be developed which, in turn,
compresses the discs between the vertebrae of the back. A rule of thumb is that the back
muscles have a moment arm of ~6 cm in relation to the L_5/S_1 disc [Jorgensen et al. 2001,

2003]. In Calculation Example 3.1, the muscle force of 2292 N may be compared with the combined gravitational force of the upper body and the box of 450 N. This means that the back muscles have to work with a contractive force which is approximately five times greater than the total gravitational force of the upper body and the box.

Calculation Example 3.1

A man weighing 80 kg and who is 180 cm tall stands bent forward holding a very light but bulky box in front of him (Figure 3.2a). The system is regarded as being in equilibrium—that is, no movement is occurring. The calculations are carried out in relation to the axis of rotation in the disc between the L_5 vertebra and the sacrum (L_5/S_1 disc). The fact that the weight of the box and the body

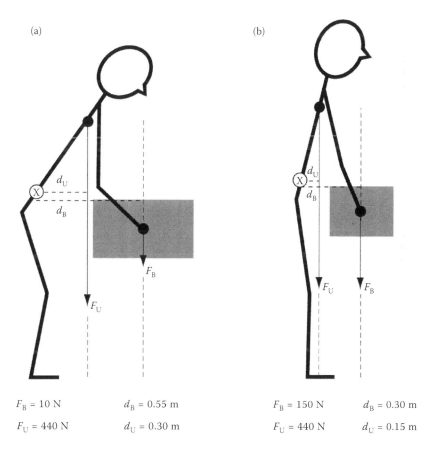

F_B = 10 N	d_B = 0.55 m	F_B = 150 N	d_B = 0.30 m
F_U = 440 N	d_U = 0.30 m	F_U = 440 N	d_U = 0.15 m

FIGURE 3.2   A man standing, leaning forward, holding a very light but bulky box in front of him (a), and standing upright holding a smaller, but heavier box in front of him (b). F_B = the gravitational force of the box in Newtons (N); F_U = the gravitational force of the upper body in Newtons (N); d_B = the moment arm of the box in relation to the L_5/S_1 disc; d_U = the moment arm of the upper body in relation to the L_5/S_1 disc. For calculations, see calculation Examples 3.1 and 3.2.

segments above the L_5/S_1 joint also contributes to a direct compression of the disc has not been taken into account.

Mass of the box: **1 kg**.
Moment arm of the box: 0.55 m.

Mass of the upper body: 44 kg.
Moment arm of the upper body: 0.3 m.

The bending-forward torque (= the external torque) on the L_5/S_1 disc:
Contribution of the box: 10 N × 0.55 m = **5.5 Nm**.
Contribution of the upper body: 440 N × 0.3 m = **132 Nm**.
Total external torque: 137.5 Nm.

Total internal torque (=extending torque): 137.5 Nm.
Moment arm of the extensor spinae muscles: 6 cm.
Contractive force in the extensor spinae muscles = compressive force
on the L_5/S_1 disc: 137.5 Nm/0.06 m = **2292 N**.

Calculation Example 3.2

The same man, as in Calculation Example 3.1, is standing upright and holding a smaller but heavier box in front of him (Figure 3.2b). The calculations have been made in the same way as in the previous calculation example.

Mass of the box: **15 kg**.
Moment arm of the box: 0.3 m.

Mass of the upper body: 44 kg.
Moment arm of the upper body: 0.15 m.

The bending-forward torque (=the external torque) on the L_5/S_1 disc:
Contribution of the box: 150 N × 0.30 m = **45 Nm**.
Contribution of the upper body: 440 N × 0.15 m = **66 Nm**.
Total external torque: 111 Nm.

Total internal torque (=extending torque): 111 Nm.
Contractive force in the extensor spinae muscles = compressive force
on the L_5/S_1 disc: 111/0.06 = **1850 N**.

The two different work situations in the calculation examples may be compared. This illustrates the relative importance of the body's contribution to the load on the lumbar spine in different work postures, and how much extra great load an external weight may add. Calculation Example 3.1 shows that the load arising from the body's own weight in an awkward work posture may represent a large proportion of the total load. In Calculation Example 3.2, the man is standing in a more favourable work posture and holding a smaller but heavier box. Here the weight of the box makes a greater contribution to the load.

The moment arm of the muscle is not constant, but varies depending on the joint angle, which means that the moment produced by one and the same muscle force changes during a movement. Inversely, a constant moment requires that muscle force varies during movement.

Lifting, in particular, has been the subject of a large number of biomechanical studies over the years [van Dieën et al. 1999]. For the most part, it is the compressive force on the L_5/S_1 disc that is being calculated, as this is regarded as a risk factor for the development of low back pain. The biomechanical models are based on simplifications and assumptions, which means that the forces calculated may be both underestimated and overestimated. Biomechanical analysis is described further in the section on methods for assessing load on the musculoskeletal system (see Section 3.11.1).

3.5.2 FACTORS AFFECTING THE SIZE OF LOAD

On the basis of biomechanical principles, a number of fundamental work factors may be distinguished that are important for the size of load on the local muscles and joints. These are factors that should be taken into account in risk assessments of different jobs and in efforts to prevent injuries as a result of heavy muscle work. In manual handling, it is primarily the following factors that are important for the size of the load on the lumbar spine:

1. The weight of the object
2. The position of the object relative to the body
3. The body weight
4. The body posture
5. The degree of symmetry/asymmetry of trunk movements
6. The velocity and acceleration of the movements

In lifting, the weight of the object (point 1) is obviously important for the size of load on the lumbar spine. If the other conditions of the lift are the same, there is a proven linear correlation between the weight of the object and the compressive force on the lumbar spine.

Points 2 and 4 are significant for the moment arms of the external forces which, according to the lever principle, are important for the size of the load. If an object being lifted is held close to the body, the moment on the lumbar spine (point 2) is reduced. The size and shape of the object affect the length of the moment arm. A large bulky object means that the moment arm is long (Figure 3.2). The work posture of the individual (point 4) affects the moment arms of both the object and the gravitational forces of the individual's own body segments (Figure 3.2). If the individual performing the lift bends the trunk forward, the moment arms of both the object and upper body weight are longer than when the individual is standing upright.

The body weight of the individual lifting also contributes to the load on the lumbar spine (point 3). In some work postures the load arising as a result of the body's own weight may represent a large proportion of the total load (see Calculation Example 3.1). Even in work situations without manual handling of objects, the load on the lumbar spine may be great if the employee is standing in an awkward work posture, that is, the work posture in itself may cause considerable load.

Additionally, the body posture of the individual is decisive for the distribution between compressive force and shear force, which acts on the back when lifting, for instance. The more the back is bent forward, the greater the shear force acting between each pair of vertebrae (Figure 4.5).

In asymmetrical movements in the trunk—that is, when the lift is performed while twisting the back—the load is particularly unfavourable on the spinal column (point 5). This is because there is tensile stress on the fibrous ring of the vertebral disc (annulus fibrosus) when the vertebrae above and below rotate in relation to each other. Tensile stress of this kind in combination with high compressive forces on the disc when lifting may cause injury to the fibrous ring. A lift may also be asymmetrical without twisting the back. When we lift or carry a load with one hand, the muscles on the opposite side from the lifting hand are activated to stabilize the back, so that we do not lean to one side. Holding a load in one hand may cause more than double the compressive force acting on the L_5/S_1 disc compared with the distribution of the load equally between both hands. When lifting with one hand during movement, these effects are, however, not as clear-cut [Cole and Grimshaw 2003].

The greater the accelerations or decelerations in a movement, the greater the muscle forces required (point 6). Lifting a box rapidly from the ground therefore requires a greater muscle force than lifting it slowly. It should be pointed out that this applies to the maximum muscle force required during a lift—that is, the peak load—and not the overall load across the entire lift, which may very well be greater when the lift is carried out slowly.

There are, of course, a number of other factors that in different ways may affect load in manual handling. Examples of factors of this kind are the ease with which the object may be grasped, requirements for precision, space, the condition of the floor (e.g., whether it is uneven, slippery, or unstable), unsuitable shoes, work pace, and staffing (i.e., to say the opportunity of cooperation). Work requiring considerable muscle force, as well as precision in work with hand-held tools, places great demands on the stability of the joints. As has been mentioned earlier, this stabilization is often achieved by muscle contraction on both sides of the joint to be stabilized (see Section 3.4). Cocontraction of this kind adds to the load on the joint. A specific problem arises when the employee does not know what a particular load weighs, or how the weight within it is distributed. Both unexpectedly light and unexpectedly heavy loads may cause problems, on the one hand, because too much effort is used and one loses balance, and on the other hand, because too little force is applied, which means that the movement is unexpectedly brought to a halt, possibly leading to overload and injury to the lumbar spine, for example.

The internal load gives rise to *physiological responses* in the various structures of the musculoskeletal system, which in turn, if one is unlucky, may give rise to injury. This will be described in subsequent sections. Of great significance for how the physiological responses develop is not only the muscle's contractive force (*amplitude*) but also the time aspect; that is, how long the muscle contracts (*duration*), how often it contracts (*frequency*), and how long it is allowed to rest and recover between each work period (Figure 1.2). Similarly, the time aspect is important when we are studying loads on other structures in the musculoskeletal system. For instance, in order to be able to assess the physiological responses to lifting objects of a certain weight, it is not sufficient to know the load level on the L_5/S_1 disc; we must also take into account how often the lift is performed during the working day, and the time between lifts.

3.5.3   WORKING TECHNIQUE

Some of the factors determining the size of load on the body—during a lift, for example—are determined by the work task itself, by how the workplace is designed and how the work is organized. The individual employee can influence other factors through the choice of *working technique* [Kjellberg 2003]. Working technique is the individual's way of "translating" an external mechanical exposure—that is, a work task—into an internal exposure (see Chapter 1, Section 1.2, Figure 1.1). Working technique may, on the one hand, be regarded as (1) the motor performance of a task, on the other, as (2) how the individual organizes the task and designs the workplace where that task is to be carried out.

The same movement or motor task may be carried out using a number of different combinations of joint movements and development of force in different muscles (see Chapter 5, Section 5.4.4). An example of variations in the motor performance is that different individuals exert themselves to different extents in performing the same work task. We can recruit a different number of muscles, use a different amount of muscle force, and activate counteracting muscles at the same time (cocontraction) to varying degrees. Another example is that we can work with moment arms of different lengths.

In lifting or in transferring a patient, the load or the patient may be held close or far away from the body. We can work slowly or quickly, smoothly or jerkily. An apparently simple task such as lifting a box from the floor up onto a table may be carried out in many different ways. One may, for example, bend one's back and knee joints to varying degrees and in different time patterns, activate different muscles and to different extents. Patient transfer is a more complicated task, and here one might expect even greater variations between individuals as to how to carry out a specific transfer. Each individual has a unique way of moving, for example, a unique gait pattern. The movement patterns of an individual presumably comprise the basis for their individual working technique.

An example of the second aspect of working technique may be that a load can be divided up into smaller portions. The working heights can be adjusted, more space created, and help provided by using a technical aid or help from a colleague. Hospital beds are often height-adjustable. The space can be adjusted by moving objects restricting the available space or by choosing to carry out the task in a place where there is sufficient space—for example, choosing the largest toilet on a hospital ward when the patient needs help moving between a wheelchair and the toilet. In patient transfer, patients can be encouraged to help out and in this way the loads to which the care staff are subjected are reduced.

There is no unanimous evidence that any specific lifting or patient transfer method is the best one to use in all lift or transfer situations. As an example, we might mention the innumerable studies comparing the stoop lift with the squat lift, where the results have been rather contradictory [van Dieën et al. 1999]. A *stoop lift* implies a lift performed by straight legs and bending the trunk forward, a *squat lift* is a lift performed by bending the knees and keeping the back straight. Both methods have their advantages and disadvantages. As regards compressive force on the lumbar spine, the squat lift is preferable if the load is lifted from a position between the feet.

If that is not possible, for example, if the weight is bulky and has to be lifted in front of the knees, the load is approximately the same size in the squat lift as in the stoop lift. The reason for this is that in this situation the squat lift does not result in a shorter moment arm than the stoop lift. As far as shear forces on the lumbar spine are concerned, these have proved to be higher in stoop lifts compared with squat lifts. Lifting using the squat lifts also requires more energy, as the entire upper body, which is lowered when we bend our knees, also has to be moved vertically. In repeated lifting, the squat lift may be felt to be more tiring, especially for the thigh muscles, which extend the knees. When people have not been taught to use a specific lifting method, they often use a technique which is a cross between the squat lift and stoop lift [Straker 2003].

The individual's choice of working technique is governed and restricted by a number of factors in the work situation, such as the character of the work task (e.g., a patient's weight and functional ability), workplace design (e.g., space), and work organization (e.g., staffing and time pressures). The choice of working technique is also governed and restricted by individual conditions such as physical capacity (including muscle strength), coordination skills, body size, experience, and knowledge [Kjellberg 2003]. The individual therefore adapts their working technique to the situation in which the task is to be carried out as well as to their own abilities.

The same individual may also vary their working technique on different occasions in performing a specific task. To what extent an individual can, for example, repeat their lifting technique in a specific lifting task depends, in part, on their experience and skill. One precondition for the individual to be able to repeat the same movement pattern and development of force is that the lifting technique has become second nature and that a motor program has been created for this skill. This means that the lift is "automated," and can be carried out without any conscious control of movements. We may compare this with techniques in sports, where the ability to repeat the movement pattern is decisive for good sporting achievements. An inexperienced beginner finds it difficult to repeat a movement pattern from time to time. This may be the reason why unnecessarily large loads sometimes occur. At the same time, a standardized working technique is not always preferable as far as load on the musculoskeletal system is concerned. If a task has to be carried out frequently, a varied movement pattern may result in the load being distributed across different structures, which reduces the risk of injury.

In manual handling work it often happens that employees receive training, for example, in lifting and transfer techniques. What characterizes a good working technique and measures to promote good working technique on the part of employees will be dealt with in Section 3.14.

3.6   GENDER ASPECTS

Different people have different preconditions, or *physical capacity* (see Chapter 1, Sections 1.2 through 1.4), to carry out physically heavy work, which also affects their physiological responses. The same internal exposure (see Chapter 1, Section 1.2 and Section 3.5) produces differing physiological responses in different employees depending on variations in their physical capacity. There are differences between women and men in their capacity to cope with work tasks requiring large muscle force, among others, differences in muscle strength and aerobic capacity (see Chapter 2) and, possibly, differences in coordination skills and motor patterns.

Women's maximal muscle strength is on average lower than that of men, which primarily results from men having larger muscle fibres and thus greater muscle mass. The difference in strength between the genders is greater in the upper extremities than in the lower. Women's maximal muscle strength in their legs is on average ~65–75% that of men, in their trunk muscles it is 60–70%, while in their elbow muscles it is only 50% [Åstrand et al. 2003, Chapter 8]. There is, however, a wide distribution in muscle strength within the two genders, which means that there are women who are stronger than men within the same age group. Women's lower maximal strength means that a job requiring certain strength, for example, lifting an object weighing 15 kg, will require a greater proportion of women's capacity compared with men's. An average woman therefore becomes more tired than an average man when lifting something.

Differences in working technique have been observed between men and women, for example, in the lifting technique [Lindbeck and Kjellberg 2001]. Differences of this kind presumably result, in part, from the fact that women have to adapt their technique and compensate for lower muscle strength in heavy manual handling work. The differences may also originate in gender differences in the flexibility of joints and muscles, anatomical differences, and possible differences in coordination skills and motor patterns. In sports, it has been shown that gender differences exist in, for example, running and throwing techniques. Differences between the genders have also been shown in the neuromuscular function and the ability of the muscles to develop force; for example, men have more rapid development of force.

The design of workplaces is sometimes better suited to certain individuals and less well suited to others. As an example, working heights may be poorly adapted for both short and tall individuals to be able to work optimally. Often, workplaces are suited to men's dimensions (anthropometrics), which results in disadvantages to women.

3.7   AGE ASPECTS

People's capacity to cope with work requiring considerable muscle force declines with increasing age; among other things, muscle strength, aerobic capacity (see Chapter 2), and certain motor functions (see below) decrease.

Maximal muscle strength is achieved generally at around the age of 20, and then declines gradually [Åstrand et al. 2003, Chapter 8]. The strength of a 65-year-old is on average 75–80% that of the strength of a person between 20 and 30 years of age. After 65 years of age, a more rapid decline in strength occurs. There are, however, major differences between individuals as to when this accelerating decline begins, which depends, among other things, on how much physical activity the individual is engaged in. The extent of the decline with age is greater in the leg and trunk muscles compared with the arm muscles. The declining muscle strength with age depends largely on the fact that the muscle mass decreases (see Chapter 6, Section 6.8).

Aging also brings with it changes in the properties of the motor units, that is to say in the neuromuscular function [Enoka 2008, Chapter 9]. Reaction time increases, the balance deteriorates, and the ability to control submaximal force (e.g., maintaining a constant grip force) declines. It is unclear as to what extent deterioration in motor ability with age is a result of aging in itself or that we engage in less physical activity with increased age.

The fact that physical activity declines with increasing age means that an imbalance may occur between the physical demands of work and the individual's physical capacity, which implies that the reserve capacity of the aging employee diminishes [de Zwart et al. 1995]. This may have the consequence that the older employee is more often subjected to loads that are too great in relation to their physical capacity, and more often has poor recovery compared with a younger employee.

One means of coping with the physical demands of work is that the ageing employee develops compensatory strategies—that is, changes their working technique [de Zwart et al. 1995]. This may be a question of using technical aids to a greater extent, enlisting the help of workmates, and working at a slower pace. How much the older employee can alter their working technique depends on how much decision latitude they have in the work situation. The experience one acquires at work with increasing age means that one becomes more skilful and efficient in carrying out the tasks, which presumably often compensates for a diminishing physical capacity. For older employees, it is particularly important to have the opportunity of exercising control over their work situation, so that they can adapt the work to their capacity and ability. Rigid control, a lack of breaks, and a high work pace may result in older people being excluded from physically heavy work.

3.8 PHYSIOLOGICAL RESPONSES TO WORK REQUIRING GREAT MUSCLE FORCE

The motor system is characterized by great adaptability [Enoka 2008, Chapters 8 and 9]. When the system is subjected to a new load, it adapts to these new demands by building up its capacity. The cells and tissues in the muscles and connective tissue, circulation system, and energy metabolism, as well as the control of movements, are all adapted. Changes are specific, that is, only the functions used in a particular physical activity, such as a particular work task, will undergo adaptation. Adaptations are also transient. As soon as a physical activity ceases, the system will adapt to the new, lower load requirements.

The mechanism underlying this adaptability consists of the constantly ongoing *remodelling process* (see also Chapter 6, Section 6.11.3, Figure 6.7). Cells and tissues in the body are continually renewed by degradation and reconstruction. The normal aging process means that the degradation is somewhat greater than the reconstruction, that is to say there is a gradual degeneration in the tissues of the body with increasing age. This ageing process is presumably governed by the genes. The degradation process proceeds at different rates in different individuals depending on genetic factors and lifestyle. The degradation is, in part, retarded and counteracted by physical activity. Subjecting oneself to mechanical load stimulates, and is necessary for, reconstruction and growth. The load must not, however, be so great that the tissues are damaged. A lack of load produces the opposite effect, that is, it stimulates a more rapid degradation of cells and tissues. In other words, both too little and too much load can weaken and damage the musculoskeletal system; the optimal load for the tissues is something in between.

With a new mechanical load, breakdown of the tissues initially takes place (the so-called *acute response*, see Chapter 1, Section 1.2 and Fact Box 3.2) (Figure 3.3). If there is sufficient time for recovery, not only does reconstruction take place, but also a reinforcement of the tissues occours (the so-called *training effect*) [Åstrand et al. 2003, Chapter 11; Enoka 2008, Chapter 9]. An adaptation of this kind presupposes that the load is greater than that to which the individual is commonly subjected, but that it is not so great as to cause damage to the tissues. The load must also be of sufficient duration, and recur at regular intervals. A training effect is more prolonged than the acute response, but is still *reversible* (so-called *long-term effect*— see Fact Box 3.2). A lack of recovery or too high a load may, on the other hand, lead to the opposite effect. The tissues are not reconstructed and micro-injuries accumulate in the tissues. If degradation of the tissues of this kind is allowed to continue for a long period, it may develop into *irreversible* damage.

The physiological responses (the training effects) of work requiring great muscle force are, under optimal conditions:

- Greater strength and endurance in the muscles.
- Greater strength in the connective tissues.
- Improved control of movements, that is, improved coordination and balance.
- Increased capacity of the circulation system and a more appropriate energy metabolism (see Chapter 2).

FACT BOX 3.2

Acute response: The immediate response from the motor system, when it is subjected to a new load on a single occasion of physical activity. The acute response to mechanical load often implies a degradation of tissues. The processes are short term and reversible, that is, they revert quickly to their original state when the load ceases.

Long-term effects: The cumulative responses from the motor system when subjected to load under recurrent physical activity for a longer time period. Under optimal conditions—that is, at the correct combination of amplitude, frequency, and duration of load—the long-term effects mean a buildup of the tissues (training effect). These adaptations are often long term, but reversible when the load ceases. Long-term effects may also, under non-optimal conditions, mean a continual degradation of tissues and impaired tolerance of new loads. The tissues become more vulnerable to injury. If this process continues for a long period, changes of this kind may become irreversible.

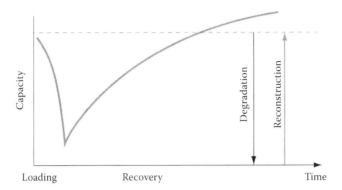

FIGURE 3.3 Degradation and reconstruction of tissues and capacity as an effect of a new load exceeding the load to which the system is usually subjected. If there is sufficient time for recovery, not only reconstruction of what has degenerated takes place, but also a reinforcement of the tissues (training effect).

3.8.1 Acute Response

Heavy muscle work may lead to fatigue in the muscles. *Muscle fatigue* may be defined as an exercise-induced impairment of the muscles' ability to produce force, and may be caused by a number of different physiological changes in the muscles [Åstrand et al. 2003, Chapter 15; Enoka 2008, Chapter 8]. Muscle contractions arise as a result of a long chain of processes, which is described in Chapter 6, Sections 6.4.4 and 6.4.5. Muscle fatigue can arise as an impaired function at each stage in this chain, both at the central level and peripherally in the muscle cells. This means that there is no single cause of muscle fatigue that may be applied to all types of work or all levels of muscle activity. Muscle fatigue is an appropriate adaptation to heavy work. The perceived fatigue is a clear signal that the muscle needs to rest so that injury does not occur. In a force development close to the maximum (100% MVC— see Fact Box 3.1), the muscle becomes fatigued within a few seconds, because the local energy substrates that power the contraction run out. When the muscle is working uninterruptedly, in what is called *static* muscle activity (see Chapter 6, Section 6.6), at ~50% of its maximal capacity (50% MVC), its endurance time is approximately 1 min. An example of work of this kind is carrying a heavy load. In uninterrupted work at 30% MVC, the endurance time is approximately 3 min. If the muscle rests between contractions—that is, in *dynamic* muscle activity—the endurance time is longer. Muscle fatigue is normally reversible, and the muscles recover during rest from work. How long this recovery takes depends on which type of work has caused the fatigue. When muscle fatigue has arisen as a result of high load over a short period, the time required for recovery is shorter compared with its having arisen over a long period of work at low load levels. Many heavy lifts one after the other lead quickly to fatigue in the muscles, but recovery also occurs relatively quickly.

After heavy muscle work which the individual is not used to, *delayed onset muscle soreness* may arise [Åstrand et al. 2003, Chapter 11; Enoka 2008, Chapter 8]. This is also a reversible process, but with a somewhat longer time horizon than for muscle fatigue. The condition is characterized by tenderness, stiffness, and weakness in the muscles, appearing approximately 24–48 h after the physical exertion and lasting approximately 3–5 days. Delayed onset muscle soreness is presumably the result of a local inflammatory reaction, but there is no generally held view among researchers as to the mechanisms behind such pains. What we do know is that it is more common that *eccentric* work (see Chapter 6, Section 6.6) triggers delayed onset muscle soreness. One cause may be mechanical damage to the connective tissue of the muscle. Muscle biopsies have also revealed damage to the structure of the muscle fibres. Delayed onset muscle soreness also constitutes a warning signal that a rest period from heavy work is necessary, so that injury does not occur in the muscles involved. If, after delayed onset muscle soreness has subsided, the work that triggered the pains is repeated a number of times, the symptoms will gradually decline.

3.8.2 LONG-TERM EFFECTS

When someone starts strength training, their *muscle strength* during the first weeks may increase by 20–40% without an increase in muscle size [Åstrand et al. 2003, Chapter 11]. This is because the initial training effect consists only of adaptations in the nervous system. The ability to activate the muscles increases, for example, through a greater synchronization of the motor units in such a way that they discharge action potentials at the same time to a greater extent (see Chapter 6, Sections 6.4.4 through 6.4.6). The synchronization of synergistically acting motor units increases, while the activity of antagonistic units decreases, that is to say, coactivation decreases. If the training continues, an adaptation also occurs in the muscles after 4–8 weeks, primarily through the cross-sectional area of the muscle fibres increasing (hypertrophy) (see Chapter 6, Section 6.4.3). The effect of strength training is specific, that is, the increase in strength is most pronounced at exactly the joint position and in the movement in which the muscle has been trained.

Heavy muscle work can also increase *muscle endurance*. In the context of training one distinguishes between strength training and endurance training. In endurance training, the ability of the muscles to make use of energy efficiently is improved. The capacity for aerobic metabolism (see Chapter 2, Section 2.4) is improved through increasing the number of mitochondria and capillaries, and thereby the ability of the muscles to use oxygen and nutrients to form ATP. The ability to make use of fat as an energy source also increases.

Heavy muscle work does not, however, always provide the training effect in muscle strength that one might expect. On the contrary, prolonged physical load at work seems to be able to reinforce the age-related degradation of muscles and other tissues (see Section 3.7). In younger people, greater muscle strength has been found in individuals with heavy manual work compared with individuals with sedentary work [Era et al. 1992; Tammelin et al. 2002], but this may result from a selection of young strong individuals to occupations of that kind. In middle-aged or older individuals, in contrast, several studies have shown that employees with heavy physical work in general seem to have lower muscle strength compared with those with a low physical load at work [Era et al. 1992; Nygard et al. 1988; Torgen et al. 1999]. Muscle endurance and physical fitness have also been shown to be worse. The absence of the training effect of heavy work may result from a monotonous, excessive load on individual muscles over a long period of time in combination with insufficient time for recovery and reconstruction. Moreover, the older employee often has a lower reserve capacity as a result of their declining muscle strength (see Section 3.7), which presumably causes the load more often to be too high and recovery often to be insufficient in comparison with a younger employee. For the work to have a constructive effect, it should presumably involve variations in the load with alternating light and heavy loads, variation in which muscles are loaded, and sufficient time for recovery between loads. In many types of work requiring great muscle force, for example, Karen's work with patient transfer, the load is presumably also too sporadic to produce a training effect. Working life seldom provides the correct combination of amplitude, frequency, and duration. There have also been discussions on whether one explanation as to why heavy work does not always result in an increased physical capacity might be that individuals with physically heavy work devote themselves to a lesser extent to physical training in their leisure time compared with individuals with sedentary work.

The load on the *skeleton* during heavy work yields a stronger and more solid skeleton. Of greatest significance is load in a longitudinal direction, that is, carrying one's own body weight and possibly other loads. The skeleton adapts to mechanical load through the load stimulating bone growth and reconstruction in the constantly ongoing remodelling process (see Chapter 6, Section 6.11.3). From 20 years of age onward, a gradual loss of bone tissue and bone minerals takes place as a result of the fact that the amount of bone tissue built up is somewhat less than the tissue that degenerates [Åstrand et al. 2003, Chapter 7]. The skeleton also changes in its composition and becomes more brittle (osteoporosis). This loss increases with advancing age, as the degradation becomes greater in relation to the new growth. The loss also becomes more rapid if the skeleton is not subjected to mechanical load.

Heavy manual work that involves the joints regularly being exposed to short-term load may have positive effects on the *joint cartilage*. Repeated short-term loads of a cyclical character stimulate the construction of cartilaginous tissue and make it harder and thicker [Åstrand et al. 2003, Chapter 7]. An acute effect of cyclical load is that the joint cartilage swells and increases in thickness, which may happen in just a few minutes. This is caused by fluid seeping into the cartilage from its surroundings when the cartilage is alternately pressed together and released (like a sponge). The increased fluid content in the joint cartilage means that the contact surfaces in a joint increase, and the compressive force per unit of area decreases, which reduces the risks of injury to the joint. One way of preventing overload of the joints in heavy work, therefore, is to warm up beforehand. The supply of nutrients to the cartilage also depends on this mechanism. In inactivity, the supply becomes insufficient, which means that the cartilage breaks down. Prolonged continuous loads, or loads that are too great, can also injure the cartilage and may lead to its degradation. Both too little or too much load thus contributes to the degeneration of joint cartilage. The changes are, however, conditioned by age and heredity to a great extent. With advanced age, the joint cartilage gradually degenerates, for example, the discs in the spinal column and the joint cartilage in the hip and knee joints. The fluid content decreases and the cartilage becomes less elastic. The degree of degeneration varies strongly between different individuals. These changes may lead to arthritic changes in the joints. It is unclear as to what extent physical load at work may affect the development of degeneration of joint cartilage and discs in the spinal column.

Tendons and *ligaments* also adapt to a greater mechanical load by becoming stronger and stiffer. Muscle work that puts strain on tendons and ligaments causes the cross-sectional area to increase and the properties of the connective tissue to change so that it becomes stronger per unit of area [Åstrand et al. 2003, Chapter 7]. Increased stiffness in a tendon increases its ability to transfer force from the muscle to the bone. In both tendons and ligaments, the ability to resist external forces increases, as well as the ability to develop force from a stretched condition. The absence of load has the opposite effect, with reduced strength and stiffness. The tendon, including its attachments to muscle and bone, has a poorer capability of adapting than the muscle, as muscle has a greater ability for metabolic activity. This may result in an imbalance in strength between the tendon and muscle, when the mechanical load increases, and in a risk of overload injuries.

The control of movements also undergoes an adaptation when we are given work tasks requiring great muscle force. The more number of times the same task is performed, the more automated the force development and motor patterns become (see Chapter 5, Section 5.4). Automation usually leads to better coordination of muscle efforts and movements, and greater economy of movement—that is, less energy is used to carry out a specific task. Automated motor patterns are difficult to change, and it is important to practise appropriate and sustainable habits from the beginning (see Section 3.5.3).

3.9 DISORDERS OF THE MUSCULOSKELETAL SYSTEM RELATED TO HEAVY MUSCLE WORK

Musculoskeletal disorders reported by employees with heavy muscle work are primarily localized in the lower back, but also in the neck, shoulders, arms, hands, hips, and knees. The acute physiological responses to mechanical load described in the previous section—that is, muscle fatigue and delayed onset muscle soreness (see Section 3.8.1)—may be seen as signals that the muscle needs rest, thus providing protection against overload. Problems in the form of discomfort, aches, and pains may also be the first signal that an injury is occurring and that something in the individual's work situation or working technique needs to be altered to avoid this. One way to prevent chronic, irreversible injury to the musculoskeletal system is, therefore, to be watchful for early signs of this kind.

3.9.1 MECHANISM OF INJURY

Most researchers agree that too great a mechanical load can give rise to injury to the various structures of the musculoskeletal system, but the exact mechanisms behind the injuries have not been clarified. Many tasks involving handling heavy objects or people give rise to loads approaching the tolerance levels of the tissues. In certain conditions, and for certain individuals, this limit is exceeded and injury occurs. Different individuals have different degrees of sensitivity to being affected by injury; their muscles, bone tissue, tendons, ligaments, and cartilage tolerate different amounts of load. How strong you are is also of great significance, of course. Differences of this kind in individuals may be genetic, or may be based on gender or age (see Sections 3.6 and 3.7, respectively) or previous exposure to mechanical load (see Section 3.8.2). The individual's working technique may, of course, also be important in terms of the risk of being injured (see Section 3.5.3).

Initially in this chapter, three types of load were described as arising during work tasks requiring great muscle force: occasional peak loads for short periods, repeated loads for a longer period, and an uninterrupted load with few breaks for rest over a longer period. The mechanisms for possible injury differ between these three types of load [McGill 1997].

An *occasional load* that exceeds tissue tolerance on one occasion is enough for an injury to occur (Figure 3.4). One example is the assistant nurse who, by herself, helps a patient weighing 90 kg to transfer from a wheelchair to a toilet in a narrow toilet space. The patient cannot help as much as the nurse had expected, but hangs onto her. At the same time the nurse has to twist her own back so they can turn around to the toilet seat. A sudden acute pain occurs in her lower back. This is usually known as a musculoskeletal injury and is a common reason for reporting occupational injuries (see Section 3.12).

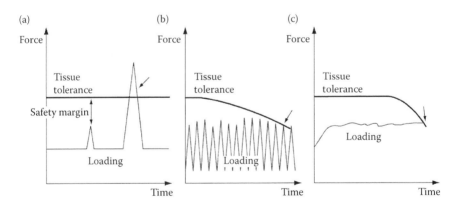

FIGURE 3.4 Different mechanisms of injury with three types of loads involved in work requiring great force: occasional high peak loads for short periods (a), repeated loads for a longer period (b), and an uninterrupted load with few breaks for rest over a long period (c). One single load on one occasion may reduce the safety margin to zero and cause injury (shown with a small arrow) if the load exceeds the tolerance level of the tissues (a). A cumulative trauma can arise with repeated loads at relatively low levels (b) or with a relatively low load continuing without interruption for a long period (c). The loads initially fall short of the tolerance level of the tissues, but this level gradually decreases. When the safety margin approaches zero, injury may occur. (Modified and reprinted from *Journal of Biomechanics*, 30(5) McGill, S.M. The biomechanics of low back injury: Implications on current practice in industry and the clinic, 465–75, Copyright (1997), with permission from Elsevier.)

Another injury scenario is a *cumulative trauma*, resulting from somewhat lower loads that are either repeated without sufficient time for recovery or that continue uninterrupted for a longer period (Figure 3.4). The repeated or uninterrupted loads initially fall below the tolerance level of the tissues, but may lead to the tissue slowly being broken down and the tolerance levels dropping. When the tolerance level has dropped so much that there is no longer any safety margin for the load to which the person is exposed, an injury occurs. One example of repeated low loads would be a warehouse worker who lifts 5 kg boxes onto pallets for a large part of his working day. Over and over he strains the tissues of his lumbar spine, which causes a slow deterioration in its tolerance level until an injury in one of the tissues occurs. As an example of an uninterrupted load with few rest periods, one might mention the postal delivery worker who carries bundles of mail weighing ~5 kg up and down flights of stairs in blocks of flats. The declining tolerance levels that appear in a cumulative trauma also increase the risk of a single overload occurring. This means that it is not always possible to distinguish these injury mechanisms from each other when an injury occurs. One example of this may be when the warehouse worker described above lifts a heavier box towards the end of the workday and feels acute pain in the lower back.

There are also other injury mechanisms concerning the way in which muscles are recruited and supplied with blood during highly repetitive and/or prolonged work with a very low force development. These mechanisms for injuries to muscles are explained in Chapters 5 and 6. It is possible that mechanisms of this kind to a certain extent might even explain how pain develops in the muscle during work requiring great muscle force.

Examples of tissue damage that may occur when the mechanical load is too high might be micro-fractures in the skeleton (e.g., in a vertebra), damage to discs (herniated disc), muscle rupture, tendon inflammation (tendinitis), ruptures in tendons and ligaments, ruptures or inflammation in joint capsules (e.g., the joint capsules in the facet joints), and degeneration of joint cartilage and discs. Micro-injuries occurring at forces below the tissue tolerance level accumulate when there is insufficient recovery, causing inflammations and degenerative processes. In low-back disorders, for instance, we often do not know exactly which structure has been damaged and what the injury looks like anatomically. This applies to both acute problems and disorders arising after a long period of load.

The degrading effects of cumulative loads may develop gradually over a very long period, causing chronic, irreversible degenerative changes, for example, in the joint cartilage and discs in the spinal column. It may be difficult to differentiate tissue changes of this kind from normal changes due to age. In this way, pain and functional impairments may arise without any specific, time-determined injury having occurred.

Altered or imperfect control of movements may contribute to injuries arising in the musculoskeletal system. The situation may arise, for instance, when one is tired or unused to a work task, after prolonged inactivity, or in cases of pain and functional impairment. Imperfect coordination or balance may entail risks of accidents in manual handling work such as improper grips, stumbling, and losing one's balance when lifting. It also happens that acute back pain occurs after a relatively simple and common back movement at low load, for example, bending down to pick up a pen that has dropped on the floor. One possible explanation is muscle control dysfunction

[McGill 1997; Panjabi 2006]. Normally, the back is stabilized during movements by the muscles between the different segments of the spine. One hypothesis that has been proposed is that injuries may arise as a result of a temporary defective function of the inter-segmental stabilizing muscles. The vertebrae may then end up in extreme postures where irritation and injury may arise in some tissue [Cholewicki and McGill 1996]. The sudden need for the body to recover stability may presumably also lead to muscle spasm and overload of individual tissues.

It is not unusual that sudden and unexpectedly high loads arise during manual handling work, such as when a patient unexpectedly falls during a transfer, or when a load that has to be lifted weighs more than expected. Such sudden loads may also disturb the stability in the spine when the CNS does not have time to resist the load by increasing the cocontraction of antagonistic trunk muscles. During patient transfers, accidents often occur when the patient behaves in an unexpected manner, such as when they suddenly resist or faint.

3.9.2 PAIN

Discomfort, aches, and pains in the musculoskeletal system do not in themselves reveal the cause of the problems. Pain in the musculoskeletal system may be caused by tissue damage or tissue irritation from any structure containing *nociceptors*— peripheral nerve endings that send pain signals when damage occurs. As an example, all the tissue components in the lumbar spine, apart from the central parts of the discs, are provided with nociceptors. Pain from the lower back may therefore be caused by an injury to the vertebra, disc, facet joint, joint capsule, ligament, muscle, blood vessel, or nerve tissue. In order for nociceptive pain to arise, it is necessary for the nociceptors to be stimulated mechanically, chemically, or thermally. Stimulation of this kind often leads to a reflex-induced muscle contraction (spasm) across the painful area, which may persist even after the initial tissue damage has healed. Another type of pain is *neurogenic pain* resulting from damage in the peripheral (e.g., pressure on a nerve root) or CNS.

Which tissues are affected in musculoskeletal disorders is often unclear, as the perception of pain is often diffusely located. Changes or deviations from the "normal" appearing on ordinary x-rays of the spine, for example, often have no connection with the individual's symptoms. Conversely, it is often difficult to find any visible changes in the tissues of patients with back pain. In these cases of so-called *non-specific disorders*, the doctor often gives a *symptomatic diagnosis*, for example, lumbago (i.e., pain located between the lowest rib and the gluteal cleft on the back of the thighs) and sciatic pain (i.e., pain with a distribution corresponding to the innervation area of the sciatic nerve). Only ~10–20% of all those seeking medical care for back pain receive a diagnosis based on the known cause. These are called *specific back disorders*, for example, disc prolapse, spinal stenosis, compression of the vertebra, and inflammatory back diseases.

The pathophysiological mechanisms behind how the majority of pain conditions in the musculoskeletal system arise are, therefore, insufficiently explained. In all probability, the physiological and psychological processes underlying the pain are complex and multi-factorial.

3.10 RISK FACTORS

From epidemiological studies, associations have been proved between a number of different exposures in working life and musculoskeletal disorders. Associations have been shown not only with both physical and psychosocial working conditions, but also with factors that do not have anything to do with work, such as individual, leisure time, and lifestyle factors. Moreover, previous musculoskeletal disorders have also been shown to be a strong risk factor for new disorders. Knowledge of how the various risk factors interact in the onset of disorders is deficient. The threshold for hazardous exposure—that is, the amplitude, frequency, and duration at which a load is harmful—is not well known. The lack of knowledge sometimes leads to the associations between work and musculoskeletal disorders being questioned, even if the situation is really that we do not know enough to be able to draw any safe conclusions.

A large number of studies have shown that *manual handling* constitutes a risk factor for *low-back disorders* [Burdorf and Sorock 1997]. More specifically, there is strong evidence that *heavy lifting* is a risk factor [Bernard 1997; da Costa and Vieira 2010; Hoogendoorn et al. 2000]. How heavy lifting is defined varies somewhat from study to study. There is a lack of studies investigating specifically how heavy, how often, and how long one can lift objects before problems arise. Frequent lifting of objects weighing at least 15 kg is often stated as a risk factor [Lotters et al. 2003]. Frequent lifting is normally defined as more than 15 lifts per day [Hoogendoorn et al. 2000]. There is some evidence for a positive association between carrying out *patient transfers* and an increased occurrence of back disorders [Burdorf and Sorock 1997].

Whole-body vibrations have been reported as a risk factor for back disorders [Bernard 1997; Burdorf and Sorock 1997]. Whole-body vibrations often occur in occupations where heavy lifting and stooped and twisted work postures are common, for example, among long-distance lorry drivers who load and unload the goods they transport. It has not been clarified as to whether whole-body vibrations alone can cause problems in the lower back, or whether they constitute a risk only in combination with other ergonomic risk factors [Okunribido et al. 2008; Palmer et al. 2003].

The relationships between work requiring considerable muscle force and *problems in the neck, shoulders, arms, and hands* are less evident. The association is clearest with regard to work with hand-held tools. For problems in the shoulder region, there is evidence for *repetitive work* with the arms held above shoulder height and work for long periods with the arms in this position [Bernard 1997; Svendsen et al. 2004; van Rijn et al. 2010]. The association becomes stronger if these postures are combined with working with hand-held tools. An association has also been shown between lifting, pushing, or pulling heavy objects and neck/shoulder disorders [Grooten et al. 2004; Harkness et al. 2003; van Rijn et al. 2010]. Using *considerable hand force* for a large part of the day, for example, by squeezing with the hand or pinching with fingers around tools, comprises a risk for disorders in the elbow, forearm and hand region, particularly for tendinitis in the elbow and wrist [Bernard 1997; Sluiter et al. 2001]. The associations are stronger if the forceful grip is combined with *repetitive movement*s of the elbow and/or wrist [Bernard 1997].

Disorders in the *hip and knee joints* connected with heavy work consist primarily of arthritis in these joints. There is consistent evidence for the fact that *heavy lifting* is a risk factor for *hip-joint arthritis* [Jensen 2008a]. There is also an established association between farm work and hip-joint arthritis [Jensen 2008a]. For *knee-joint arthritis*, there is evidence for heavy lifting and work in a kneeling and squatting position [Jensen 2008b; McMillan and Nichols 2005]. It is difficult to separate the effects of lifts from the effects of work postures loading the knees, as these exposures often occur in the same occupations, for example, among floor-layers and miners.

3.11   METHODS FOR ASSESSING LOAD ON THE MUSCULOSKELETAL SYSTEM

3.11.1   BIOMECHANICAL MODELS

Biomechanical models may be used to calculate the load on the musculoskeletal system in work tasks requiring great forces. Calculations are based on measurements of external forces, body postures, and movements over time. Biomechanical models have varying degrees of complexity. A *static analysis* is the simplest and most elementary means of assessing load on the musculoskeletal system. In an analysis of this kind, the fact that a movement is occurring is not taken into account, but an object or system is studied as if it is at rest or at movement equilibrium. If we use a static model to calculate forces on the muscles and joints in lifting, then we "freeze" the movements and calculate the forces at a specific body posture, for example, in the starting position when the load is just about to be raised from the ground. We do not take into account the fact that accelerations take place at the same time. What is necessary is to measure the angles of the joints. In the simplest analysis, the only external forces taken into account are the gravitational forces (from the load being lifted and from one's own body parts). Calculation Examples 3.1 and 3.2 are examples of static analyses of this kind.

A truer picture of the loads arising during movements is given by a *dynamic analysis*. When analysing rapid movements, it is necessary to use a dynamic analysis. As an example, the peak load on the lumbar spine may become twice as great when a load is lifted quickly compared with holding it still. If a static analysis is performed on such a rapid lift the load may be seriously underestimated. In a dynamic analysis, the system is regarded as being in motion and subjected to forces that cause accelerations. For these calculations, it is necessary to measure the accelerations of the movements. In manual handling work, the hands are often exposed to forces other than gravitational forces alone, for example, when pushing or pulling an object, or working with a machine or hand tool. Here it is necessary to measure the size and direction of the forces from, or applied to, the object, by *force transducers* or dynamometers, and how the forces change during the work cycle being analysed. Alternatively, we may measure the reaction forces from the ground, the ground reaction force, with a so-called force plate.

The models may be either two dimensional or three dimensional. The simplest and often the most practical applicable analysis is the *two-dimensional* one. In Figure 3.2, the work is studied in one plane, that is to say, it is assumed that it is carried out without twisting the trunk. In order to analyse work tasks in which movements occur in several planes, *three-dimensional analyses* may need to be carried out. In patient transfers, the health care provider's movements and exertion of force rarely take place solely in one plane. An example is when a patient is to be transferred from the edge of the bed into a wheelchair. Even if asymmetrical movements are involved in most work situations, a two-dimensional analysis may often provide a good idea of the size of the load.

The most sophisticated biomechanical models are the dynamic three-dimensional models. Technical developments in the computer field have made it possible to carry out such comprehensive analyses in a short time. Collecting data for these models is, however, still very time consuming and can mostly only be done in the laboratory.

The biomechanical analyses do not usually take into account the fact that there are often stabilizing cocontractions, which add to the load. This is one reason why the forces that exist "in reality" are presumably considerably greater than those calculated using biomechanical models. It is also the case that different people may use different strategies for recruiting muscles to produce the same force, which biomechanical analysis does not take into account. The models also provide deficient information about forces acting on individual tissues.

The time aspect is of great significance for how a load affects the physiological responses in the tissues, and whether the load gives rise to injuries. The biomechanical calculation only gives a snapshot of the load and gives no guidance as to how tiring a job is, or how great the cumulative load is during a work shift.

3.11.2 ELECTROMYOGRAPHY

Electromyography (EMG) is a method for directly measuring the muscle activity level, which reflects muscle force development. Electrodes register the electrical signals generated in the muscle when it contracts, known as action potentials (see Chapter 6, Sections 6.4.5 and 6.12.6).

EMG can be used to estimate the force development in the muscle. When the intensity of the muscle contraction increases, EMG activity also increases. This relationship varies from muscle to muscle, from individual to individual, and from one measurement occasion to another. In order to be able to compare the activity from different measurement occasions, the EMG activity is often expressed in relation to the activity in well-defined test contractions, for example, an MVC (MVC—see Fact Box 3.1). The EMG amplitude at work may then be expressed as %MVC. Analysing the amplitude of the EMG signal as a measure of muscle load is an important and common method in studies of load on the musculoskeletal system at work. Analyses of this kind are simplest to interpret in work carried out with slow movements with limited ranges of motion. In rapid movements with considerable ranges of motion it is more difficult to know what the EMG signal represents. The analyses may be used to compare different methods of carrying out manual handling work, for example, in using different technical aids, or in evaluating changes in workplace design. EMG registration can be carried out throughout the entire working day, which means that we also obtain a measure of the cumulative load of an occupation.

EMG can also be used to study the coordination between muscles in movement (e.g., a work task) by registering which muscles are involved and when these muscles are active.

3.11.3 MEASUREMENTS OF EXTERNAL EXPOSURE

The load on the musculoskeletal system may also be estimated from measurements of external exposure (see Section 3.5). The size and direction of external forces may be measured with spring dynamometers and electronic *force transducers*. Information about the weight of an object being lifted may also be used to assess the internal load. Body postures (joint angles) may be recorded with measuring devices, for example, inclinometers, accelerometers, and *video based measurement systems* [David 2005]. With these instruments, measurements can be made throughout entire working days. It can also be observed how often and how long the employee adopts specific work postures, how often and how long he or she lifts or carries out some other activity, and the weight of the load lifted. A large number of observational methods are available [Takala et al. 2010]. Trained observers can carry out *observations* of this kind with a stopwatch, paper, and pen. The observations can also be registered directly on small hand-held computers with the help of specially developed software. The observations may be made directly at the workplace or indirectly by first videotaping the work to be observed. The advantage of video observations is that the recorded sequences may be observed several times, and in this way more variables may be registered than in direct observations. Moreover, employees can themselves report which activities are being carried out, how long they last and how often. *Self-reporting* of joint angles, for example, has proved to have low reliability, however. With the aid of *diaries*, repeated registrations may be carried out for a long period.

3.11.4 MODELS AND CHECKLISTS

There are more or less simple models and checklists for directly identifying and assessing work situations with hazardous loads on the musculoskeletal system. In the provisions of the Swedish National Board of Occupational Safety and Health on Ergonomics for the prevention of musculoskeletal disorders [SWEA 1998], there are simple so-called three-zone models (red–yellow–green) for assessments of lifting (see Section 3.13) and pushing and pulling work [SWEA 1998, Appendix A]. In 1981, the US National Institute for Occupational Safety and Health (NIOSH) presented its *Work Practices Guide for Manual Lifting*, recommendations for lifts based on psychophysical, energetic, biomechanical, and epidemiological criteria [NIOSH 1981]. These NIOSH guidelines were subsequently revised in 1991 [Waters et al. 1994]. The lifting recommendations are formulated as a relatively simple function in which the weight limits for symmetrical two-handed lifts may be calculated. The calculations are based on a maximal weight of 23 kg that can be lifted without risk on a few occasions under optimal conditions. This "permissible" weight is reduced by six multiplication factors: horizontal distance, vertical height, vertical lift distance, number of lifts per minute, degree of asymmetry, and ease of grip. The recommended weight limit represents a weight which almost all healthy workers can manage to lift without involving a risk of developing low-back pain.

3.12 PREVALENCE OF MUSCULOSKELETAL DISORDERS AS A RESULT OF HEAVY MUSCLE WORK

Work-related disorders in the musculoskeletal system are more common in occupations characterized by heavy manual handling than the average among women and men in employment. European statistics show that 25% of the workers in Europe report back problems [Eurofound 2007]. In Sweden, between 2005 and 2010, just over a fifth of the women working as cooks and postal delivery workers reported that they had problems as a result of heavy manual handling, compared with 6% of all women in employment [SWEA 2010b]. Among men, 22% of concrete workers, 19% of bricklayers, and 16% of plumbers stated that they had problems caused by heavy manual handling, compared with 5% of all men in employment. In many of these occupations back disorders are the most frequent kind of problem, while in other occupations in which considerable muscle force is required, problems are also common in the neck, shoulders, and arms.

In the United Kingdom, more than one-third of all musculoskeletal workplace injuries reported each year are caused by manual handling [HSE 2009]. In Sweden, the Swedish Work Environment Authority annually publishes statistics on occupational injuries based on the accidents at work and work-related disorders reported to the Swedish Social Insurance Administration. Of those accidents at work that resulted in absence in 2005, physical overload was given as the cause of the accident (called musculoskeletal injury) in just over one in four of the accidents reported among women and just over one in six of those reported among men [SWEA 2007]. Among women, half of these musculoskeletal injuries had occurred in contact with people, for example, lifting and transferring people. Assistant nurses, nurse's aids, care assistants, and personal assistants are the female occupational groups reporting the most musculoskeletal injuries. Six out of every 1000 employees reported a musculoskeletal injury between 2004 and 2006. For men, the most vulnerable occupational group is fire fighters, of which 11 out of every 1000 employees reported a musculoskeletal injury during this time period. For the entire working population, the average was 1.6 cases per 1000 employees. Of the work-related disorders reported, six out of 10 were caused by load factors (called musculoskeletal disorders) for both genders [SWEA 2007]. An injury is assessed as being a musculoskeletal disorder if it has arisen through the effect over a long period of heavy lifting, or monotonous and awkward work postures and work movements. In just over half of all musculoskeletal disorders reported by both men and women, lifting and transferring heavy loads was stated to have contributed to the onset of the injury.

3.13 WHAT THE LAW SAYS ABOUT WORK REQUIRING CONSIDERABLE MUSCLE FORCE

Within the EU there is a general framework directive concerning measures to promote improvements in employee safety and health at work [EUR-Lex 1989]. This directive regulates employers' responsibilities for ensuring that employees are able to carry out their work without risk to their health. The employer is obliged to:

- Adapt the work to the individual, especially as regards the design of the workplace, the choice of work equipment, and the choice of working and production methods.
- Develop a coherent overall prevention policy that covers technology, organization of work, working conditions, and the influence of factors related to the working environment.
- Give appropriate instructions to the workers [EUR-Lex 1989, article 6].

The directive also states that the employer shall ensure that each worker receives adequate health and safety training, in particular in the form of information and instructions specific to his workplace or job [EUR-Lex 1989, article 12]. The training shall be carried out on recruitment, in the event of a transfer or a change of job, in the event of the introduction of new work equipment or a change in equipment, and in the event of the introduction of any new technology. The training shall be adapted to take account of new or changed risks, and repeated periodically if necessary.

The obligations of the worker are also regulated in the directive [EUR-Lex 1989, article 13]. The worker is responsible for following the instructions given by the employer, taking care of their own health and safety and that of coworkers in accordance with the instructions and their training, and making correct use of necessary equipment.

Within the EU there is also a specific minimum directive for manual handling [EUR-Lex 1990]. Manual handling of loads is defined in this directive as "any transporting or supporting of a load, by one or more workers, including lifting, putting down, pushing, pulling, carrying or moving of a load, which, by reason of its characteristics or of unfavourable ergonomic conditions, involves a risk particularly of back injury to workers" [EUR-Lex 1990, article 12].

The directive takes as its starting point the fact that employers are instructed to attempt to avoid manual handling of loads in all circumstances [EUR-Lex 1990, article 3]. To avoid the need for manual handling, the employer shall take appropriate organizational measures, or shall use the appropriate means, in particular mechanical equipment. If this cannot be avoided, the employer shall take other measures to minimize the risks of this work. In an appendix to the directive, there are a large number of factors and aspects that have to be taken into account—for example, the characteristics of the load, the characteristics of the work environment, whether the lift is carried out with the trunk twisted, how long and how often lifts have to be performed, and individual risk factors [EUR-Lex 1990, Appendix 1]. The employer shall organize the workplace in such a way as to make such handling as safe and healthy as possible [EUR-Lex 1990, article 4].

Moreover, the employers must ensure that workers receive proper training in working technique and information about the possible risks that exist, and how they can be avoided [EUR-Lex 1990, article 6]. The employer must also make sure that the worker receives precise information on the weight of the load and the centre of gravity of the heaviest side when a package is eccentrically loaded.

Another important requirement in the directive concerns participation of employees and their representatives, for example, those representatives with specific responsibility for the safety and health of workers, in questions relating to safety and health in manual handling work [EUR-Lex 1990, article 7].

Many countries have more detailed national rules—provisions, for instance—governing this area. There exist national guidelines and general advice as a support for measures to prevent risks at work. As an example, in the provisions of the Swedish National Board of Occupational Safety and Health on Ergonomics for the prevention of musculoskeletal disorders [SWEA 1998], manual handling, and other exertions of force are dealt with in a specific section (Section 3). It states that the employer shall ensure that manual handling and other work requiring exertion of force shall as far as is practicable be organized and designed in such a way that the employee can work without being exposed to physical loads that are injurious to health or unnecessarily fatiguing. In the comments to this paragraph, one further step is taken, where it is stated that employers shall in the first place investigate whether manual handling can be avoided completely. If this is not possible, risk analysis of the work should be carried out, and measures taken subsequently, for example, regarding the design of loads, provision of technical aids, and work organization measures. It is also stated that normally, manual lifting of people should not need to be carried out within the health and social care. It is possible to transfer people without lifting them.

No absolute threshold is given for the weight of a load to be lifted, but it is said that many factors affect the risk of injury in a lifting situation. In a model for assessing lifts, two factors are taken into account: the weight of the load to be lifted and the distance between the load and the body [SWEA 1998, Appendix A]. The model applies to symmetrical lifts with two hands under ideal conditions.

What is recommended here is:

- Maximum 25 kg when the load is within forearm distance (~30 cm) of the body.
- Maximum 15 kg when the load is within three-quarter arm distance (~45 cm) of the body.

The model has a list of influencing factors that should be taken into account, for example, how long and how often lifts have to be carried out, whether the lift is carried out with a stooped or twisted body, whether the object is difficult to grasp, and whether the person lifting is strong or weak. The more the number of "aggravating" factors, the more one should reduce the maximum weight given in the model.

3.14 WHAT CAN BE DONE TO REDUCE THE RISKS OF HEAVY MUSCLE WORK?

To reduce the risks of heavy muscle work, a number of measures have to be taken in many different areas. Many of these measures are touched upon in the European legislation relating to this area (see Section 3.13).

3.14.1 Measures at the Workplace

3.14.1.1 The Work Task

Manual handling of heavy loads should as far as possible be avoided entirely. When heavy lifting is necessary, *technical aids* should be used. Technical aids mean everything from machinery to trucks, hoists, and carts. Employees can push or pull loads instead of lifting them. In pushing and pulling work, it is important to have low friction between the load and the ground, but high friction between the shoes of the employees and the ground. Employees should roll their loads instead of carrying them. When running trolleys and patients' beds, for example, it should be possible to run these with reasonable strength. The wheels should be easy to steer and provide low friction against the ground. The trolley should be provided with handles at a suitable height.

Loads can be divided into smaller, lighter units. It is important to acquire information about the weight of the load, and how the mass is distributed within it, in order to be able to adapt one's effort and working technique.

Within nursing care, a common policy is that patients should not be lifted manually; that is, that one should not lift the entire body weight of the patient from the ground. If the patient cannot participate in the transfer, a mechanical hoist or other transfer aids should be used. Most patient transfers can be carried out using technical aids. When transferring a patient in the bed, the patient can be pulled instead of lifted, for example, with the aid of a draw-sheet. The participation of the patient will facilitate a transfer. This presupposes that the patient is informed about how the transfer is to be carried out. A common cause of musculoskeletal injuries among care staff is the patient suddenly resisting or behaving in an unexpected manner. Another assumption is that the transfer technique is adapted to the patient's preconditions and ability. For technical aids to be useful they have to be easily available, functional, and suited to the purpose. Staff must have adequate knowledge of how they are used.

In work with hand-held machines, tools, and controls, these should be designed so that they allow a grip that is adjusted to the requirements for both power and precision, that the gripping force is evenly distributed across the hand, and that the grip allows for a neutral position of the hand and wrist. They should suit the hand sizes of different users. Moreover, they should not require too great a trigger force, be as light as possible and well balanced.

3.14.1.2 Workplace Design

The workplace should be designed so that there is sufficient *space* for work in suitable work postures, space to use technical aids and sufficient for two coworkers to cooperate. Within nursing care, both in an institution and in private accommodation, it is unsuitable for beds to be placed against a wall if the patient needs assistance during transfers. There should be sufficient space around the bed for two care staff to collaborate. The bed should also be height-adjustable to allow safe work postures for the care staff. There should be sufficient space around the toilets. If it is not possible to arrange sufficient space in a toilet, one should consider whether the transfer could take place somewhere else. This may be arranged so that the transfer takes place outside the toilet onto a mobile toilet seat which is then rolled into the toilet.

Some thought should be given to where objects that have to be transferred are located, both as regards their starting point and their final position. The distance of the transfer should be as short as possible. If it can be avoided, the objects to be lifted should not be placed at floor level and should not be lifted to positions above shoulder height.

Stairs should be avoided as a transport route for heavy or cumbersome loads. Consideration should be given to managing obstacles in the form of differences in level, such as doorsteps, or slippery, uneven, sloping, or unstable ground.

3.14.1.3 Organization of Work

Work involving elements of heavy manual handling should also contain lighter work tasks, so that there is an opportunity for recovery. The heavier the load to be handled, the more time should elapse between lifts.

In heavy manual handling, for example, in patient transfers, there should always be an opportunity of working together with coworkers. This presupposes good staff planning and adequate *staffing levels*. How the work is organized has great significance, such as whether the organization of the nursing care on a hospital ward allows cooperation, or whether workers in the home care service work alone or in pairs.

How much *time* is allocated for a work task affects how it is carried out. If a worker feels under time pressure, they may perhaps not take the time to fetch an aid, or to go and ask a colleague for help. This may also result in "carelessness" in how the worker performs the task, and in working with rapid and jerky movements.

One way of increasing the prerequisites for employees to use safe working techniques and make use of aids is to introduce a policy at the workplace on how various work tasks should be carried out. On a ward, as an example, a policy of this kind would cover how many people should assist, as well as which transfer methods and transfer aids should be used in patient transfers. It may be a good idea to document how each individual patient should be transferred, so that all the staff use the same method. There should also be training routines for working technique (see Section 3.14.2).

One overarching aspect of the organization of work is how great an *influence* the employees have over their own work. This applies, for example, to what should be done, when and where tasks should be carried out, and working methods for these. Increasing the employees' decision latitude is one way of making it possible for the individual to be able to adapt their work to their own capacity, and in this way reduces the risk of overload. The fact that the work must be adapted to the individual is regulated in the EU's general framework directive, which concerns measures to promote improvements in employee safety and health at work [EUR-Lex 1989] (see Section 3.13). In Sweden, the employer has a duty prescribed by law to give the employee an opportunity of influencing the planning and execution of their own work [SWEA 1998].

3.14.2 MEASURES AT THE INDIVIDUAL LEVEL

One precondition for employees to be able to carry out their work in a suitable way and to avoid hazardous loads is that they have good skills in *working technique*. The employer is responsible for ensuring that the employee has the training necessary to

be able to prevent musculoskeletal disorders (see Section 3.13). It is important to practise safe working technique from the beginning. It may be difficult later to alter
ingrained motor patterns. A good starting point is to undergo *training in working
technique*, such as lifting and transfer technique. The principles of safe working technique in lifting and in patient transfers may be studied in Fact Boxes 3.3 and 3.4. Most
of these principles can be derived from the biomechanical factors affecting the size of
the load on the musculoskeletal system (see Section 3.5.2). The training should be
focused on encouraging the participants, on the basis of these principles, to adapt
their working technique to the specific work situation and to their own and the
patient's abilities. Working technique training should be included in any introduction
course for new employees. It is important that the training is long enough and that it
is repeated regularly. It is also important that it is adapted to the conditions at the
employee's own workplace, and that it is supported by the organization through ensuring that there is space for the employee to actually use safe working technique. Apart
from training courses, there needs to be opportunities for training and instruction at
the workplace. Having good role models among colleagues may also be important.

The principles of lifting technique (Fact Box 3.3) may be applied to most manual
handling tasks, even to patient transfers. Apart from these, there are a number of
more specific principles applying to the special situation represented by a patient
transfer (see Fact Box 3.4).

FACT BOX 3.3

Principles of safe working technique in lifting:

- Reduce the weight of the load.
- Lift the load close to the body.
- Avoid lifting below knee level or above shoulder level.
- Avoid lifting from a starting position where the trunk is extremely
  bent forward.
- Avoid lifting while at the same time twisting the back.
- Avoid jerky movements.
- Use technical aids in heavy lifts.
- Cooperate with colleagues in making heavy lifts.

FACT BOX 3.4

Principles of safe working technique during patient transfers:

- Inform the patient.
- Allow the patient to perform as much as possible of the transfer.
- Use transfer aids.
- Adjust the height of the bed.

- In most cases there should be at least two employees cooperating.
- Never lift the entire weight of the patient from the base—use a mechanical hoist.
- Pull instead of lifting.

In patient transfer, care should also be taken to make the transfer comfortable and safe for the patient. Moreover, it is a common dilemma within nursing care that the task of the staff is to rehabilitate patients, which means that patients must be trained to carry out, as much as possible, the transfer themselves. This may result in not using transfer aids, despite the fact there is a risk that the load on the care staff will be too great.

In order to cope with tasks requiring great muscle force, it is necessary for the employee to have a sufficiently high physical capacity. As mentioned earlier, the effect on the muscles and other tissues provided by load at work is often insufficient to build them up and increase the individual's physical capacity (see Section 3.8.2). This applies especially to the older employee [Ilmarinen 2001]. Employees carrying out heavy muscle work need to have a greater physical capacity, including muscle strength, than is usually required at work to be able to manage occasional peak loads. An example is Karen's work with patient transfers, where she is subjected to extra large loads when a transfer goes wrong, such as when a patient faints during the transfer. It is therefore recommended that employees with this type of work carry out some *physical training*. Specific training of *muscle strength* and *endurance* are important. In Karen's work, what is needed are primarily back, abdominal, shoulder, and leg muscles that are strong and have sufficient endurance to manage the heavy patient transfers. This applies to most types of manual handling work. It is also important to receive all-round training of muscle strength, muscle endurance, fitness, mobility, balance, and coordination. Good *physical fitness* is important to increase endurance and thereby reduce fatigue at work. Often this type of work also makes great demands on energy metabolism. Fatigue can lead to impaired control over movements and carelessness, which may result in accidents at work, such as tripping or grasping a load in an incorrect way. Good *mobility*, *balance*, and *coordination* are also significant for adequate control of movements and for the ability to have good working technique.

Studies have shown that physical training can alleviate and speed up recovery from back problems [Professional Associations for Physical Activity (Sweden) 2010]. Training strength, endurance, mobility, as well as physical fitness, has a positive effect. Studies on the preventive effect from physical activity on back pain do not, however, show consistent results.

Providing employees with physical training in working hours may be one way for the employer to help ensure that they have sufficient physical capacity to cope with their tasks and maintain their work ability right up to retirement.

3.15 SUMMARY

Work requiring considerable muscle force often involves heavy tasks that last only a short period of time, but are repeated a number of times during the working day. The muscles are activated close to their maximal capacity for a short period, and then later relaxed. This type of work is common in heavy manual handling; that is, lifting and moving loads such as in the nursing and care professions and in construction and warehouse work. In occupations involving manual handling, uninterrupted work requiring somewhat lower muscle force for a longer period is also common—holding and carrying loads, for example. Manual handling involves muscles in the trunk, legs, shoulders, and arms. Another common example of work requiring great force is work with hand-held tools, for example, among plumbers and carpenters, which in particular puts strain on the muscles of the hand and forearm. The force the muscles need to produce to move a load depends primarily on the weight of the load, how far from the body the load is held, the work posture of the individual, and how quickly the load is lifted. In an optimal work situation, the individual can influence these factors through their choice of working technique. Using a careful working technique may be a method of avoiding harmful loads on the muscles, tendons, and joints. The size of the load arising on these structures may be calculated by using biomechanical methods. Different individuals have different physical preconditions for carrying out physically heavy work. The lower maximal muscle strength of older individuals and women compared with younger individuals and men means that a specific job will require a greater proportion of their capacity. Under optimal working conditions—that is, at the correct combination of amplitude, frequency, and duration of mechanical load during work—an adaptation of, among other things, muscle strength and tissue strength to the requirements of the work takes place. On the other hand, a lack of recovery or too high a load may cause injuries. Musculoskeletal disorders are more common in occupations characterized by heavy manual handling than the average among working women and men. Manual handling of heavy loads should as far as possible be avoided entirely. For example, patient transfers in the nursing and care professions should be carried out using methods other than lifting. In European legislation, the employer is instructed to avoid heavy manual handling as far as possible. Where the need for the manual handling of loads by workers cannot be avoided, the employer shall take the appropriate measures to reduce the risks of manual handling, for example, by providing technical aids and ensuring that the staff regularly receives training in working technique. Heavy work should be supplemented by physical training, as the load provided by the work is rarely constituted in such a way that muscles and tissues are built up to a sufficient extent.

REFERENCES

Åstrand, P.-O., K. Rodahl, H. Dahl, and S. Strömme. 2003. *Textbook of Work Physiology. Physiological Bases of Exercise.* 4th ed. Champaign: Human Kinetics.

Bernard, B.P., ed. 1997. *Musculoskeletal Disorders and Workplace Factors. A Critical Review of Epidemiological Evidence for Work-Related Musculuskeletal Disorders of the Neck, Upper Extremity and Low Back.* Cincinnati: CDC-NIOSH.

Burdorf, A. and G. Sorock. 1997. Positive and negative evidence of risk factors for back disorders. *Scand J Work Environ Health* 23:243–256.

Cholewicki, J. and S.M. McGill. 1996. Mechanical stability of the *in vivo* lumbar spine: Implications for injury and chronic low back pain. *Clin Biomech* 11(1):1–15.

da Costa, B.R. and E.R. Vieira. 2010. Risk factors for work-related musculoskeletal disorders: A systematic review of recent longitudinal studies. *Am J Ind Med* 53(3):285–323.

Cole, M.H. and P.N. Grimshaw. 2003. Low back pain and lifting: A review of epidemiology and aetiology. *Work* 21(2):173–84.

David, G.C. 2005. Ergonomic methods for assessing exposure to risk factors for work-related musculoskeletal disorders. *Occup Med (Lond)* 55(3):190–9.

van Dieën, J.H., M.J.M. Hoozemans, and H.M. Toussaint. 1999. Stoop or squat: A review of biomechanical studies on lifting technique. *Clin Biomech* 14(10):685–96.

Enoka, R.M. 2008. *Neuromechanics of Human Movement*. 4th ed. Champaign: Human Kinetics.

Era, P., A.L. Lyyra, J.T. Viitasalo, and E. Heikkinen. 1992. Determinants of isometric muscle strength in men of different ages. *Eur J Appl Physiol Occup Physiol* 64(1):84–91.

EUR-Lex. 1989. Council Directive 89/391/EEC of 12 June 1989 on the introduction of measures to encourage improvements in the safety and health of workers at work. http://eur-lex.europa.eu/LexUriServ/LexUriServ.do?uri=CELEX:31989L0391:EN:HTML

EUR-Lex. 1990. Council Directive 90/269/EEC of 29 May 1990 on the minimum health and safety requirements for the manual handling of loads where there is a risk particularly of back injury to workers. http://eur-lex.europa.eu/LexUriServ/LexUriServ.do?uri=CELEX:31990L0269:EN:HTML

Eurofound. 2007. Fourth European Working Conditions Survey. European foundation for the improvement of the living and working conditions. http://www.eurofound.europa.eu/pubdocs/2006/98/en/2/ef0698en.pdf

Eurofound. 2010. Fifth European Working Conditions survey—2010. European foundation for the improvement of the living and working conditions. http://www.eurofound.europa.eu/surveys/ewcs/2010/index.htm

Grooten, W.J., C. Wiktorin, L. Norrman et al. 2004. Seeking care for neck/shoulder pain: A prospective study of work-related risk factors in a healthy population. *J Occup Environ Med* 46(2):138–46.

Harkness, E.F., G.J. Macfarlane, E.S. Nahit, A.J. Silman, and J. McBeth. 2003. Mechanical and psychosocial factors predict new onset shoulder pain: A prospective cohort study of newly employed workers. *Occup Environ Med* 60(11):850–7.

Hoogendoorn, W.E., P.M. Bongers, H.C. de Vet et al. 2000. Flexion and rotation of the trunk and lifting at work are risk factors for low back pain: Results of a prospective cohort study. *Spine* 25(23):3087–92.

HSE. 2009. *Getting to Grips with Manual Handling. A Short Guide*. London: Health and Safety Executive.

Ilmarinen, J.E. 2001. Aging workers. *Occup Environ Med* 58(8):546–52.

Jensen, L.K. 2008a. Hip osteoarthritis: Influence of work with heavy lifting, climbing stairs or ladders, or combining kneeling/squatting with heavy lifting. *Occup Environ Med* 65(1):6–19.

Jensen, L.K. 2008b. Knee osteoarthritis: Influence of work involving heavy lifting, kneeling, climbing stairs or ladders, or kneeling/squatting combined with heavy lifting. *Occup Environ Med* 65(2):72–89.

Jorgensen, M.J., W.S. Marras, K.P. Granata, and J.W. Wiand. 2001. MRI-derived moment-arms of the female and male spine loading muscles. *Clin Biomech* 16(3):182–93.

Jorgensen, M.J., W.S. Marras, P. Gupta, and T.R. Waters. 2003. Effect of torso flexion on the lumbar torso extensor muscle sagittal plane moment arms. *Spine J* 3(5):363–9.

Kjellberg, K. 2003. Work technique in lifting and patient transfer tasks. Doctoral thesis, Arbete och Hälsa, 2003:12, Institute of Internal Medicine, Department of Occupational Medicine, The Sahlgrenska Academy at Göteborg University, National Institute for Working Life, Göteborg.

Lindbeck, L. and K. Kjellberg. 2001. Gender differences in lifting technique. *Ergonomics* 44(2):202–214.

Lotters, F., A. Burdorf, J. Kuiper, and H. Miedema. 2003. Model for the work-relatedness of low-back pain. *Scand J Work Environ Health* 29(6):431–40.

McGill, S.M. 1997. The biomechanics of low back injury: Implications on current practice in industry and the clinic. *J Biomech* 30(5):465–75.

McMillan, G. and L. Nichols. 2005. Osteoarthritis and meniscus disorders of the knee as occupational diseases of miners. *Occup Environ Med* 62(8):567–75.

NIOSH. 1981. *Work Practices Guide for Manual Lifting.* Cincinnati, Ohio: National Institute for Occupational Safety and Health.

Nygard, C.H., T. Luopajarvi, T. Suurnakki, and J. Ilmarinen. 1988. Muscle strength and muscle endurance of middle-aged women and men associated to type, duration and intensity of muscular load at work. *Int Arch Occup Environ Health* 60(4):291–7.

Okunribido, O.O., M. Magnusson, and M.H. Pope. 2008. The role of whole body vibration, posture and manual materials handling as risk factors for low back pain in occupational drivers. *Ergonomics* 51(3):308–29.

Palmer, K.T., M.J. Griffin, H.E. Syddall et al. 2003. The relative importance of whole body vibration and occupational lifting as risk factors for low-back pain. *Occup Environ Med* 60(10):715–21.

Panjabi, M.M. 2006. A hypothesis of chronic back pain: Ligament subfailure injuries lead to muscle control dysfunction. *Eur Spine J* 15(5):668–76.

Professional Associations for Physical Activity (Sweden). 2010. *Physical Activity in the Prevention and Treatment of Disease.* Stockholm: Swedish National Institute of Public Health.

van Rijn, R.M., B.M. Huisstede, B.W. Koes, and A. Burdorf. 2010. Associations between work-related factors and specific disorders of the shoulder—A systematic review of the literature. *Scand J Work Environ Health* 36(3):189–201.

Sluiter, J.K., K.M. Rest, and M.H. Frings-Dresen. 2001. Criteria document for evaluating the work-relatedness of upper-extremity musculoskeletal disorders. *Scand J Work Environ Health* 27(Suppl 1):1–102.

Statistics Sweden. *Statistical database, The Work Environment.* http://www.ssd.scb.se/databaser/makro/start.asp (accessed 110103).

Straker, L. 2003. Evidence to support using squat, semi-squat and stoop techniques to lift low-lying objects. *Int J Ind Erg* 31(3):149–60.

Svendsen, S.W., J.P. Bonde, S.E. Mathiassen, K. Stengaard-Pedersen, and L.H. Frich. 2004. Work related shoulder disorders: Quantitative exposure-response relations with reference to arm posture. *Occup Environ Med* 61(10):844–53.

SWEA. 1998. *Ergonomics for the Prevention of Musculoskeletal Disorders,* AFS 1998:1. Stockholm: Swedish National Board of Occupational Safety and Health.

SWEA. 2007. Occupational accidents and work-related diseases 2005 (Arbetsskador 2005) (In Swedish). In *Arbetsmiljöstatistik.* Stockholm: Swedish Work Environment Authority.

SWEA. 2010a. *The Work Environment 2009.* Stockholm: Swedish Work Environment Authority. http://www.av.se/dokument/statistik/officiell_stat/ARBMIL2009.pdf

SWEA. 2010b. *Work-Related Disorders 2010.* Stockholm: Swedish Work Environment Authority. http://www.av.se/dokument/statistik/officiell_stat/ARBORS2010.pdf

Takala, E.P., I. Pehkonen, M. Forsman et al. 2010. Systematic evaluation of observational methods assessing biomechanical exposures at work. *Scand J Work Environ Health* 36(1):3–24.

Tammelin, T., S. Nayha, H. Rintamaki, and P. Zitting. 2002. Occupational physical activity is related to physical fitness in young workers. *Med Sci Sports Exerc* 34(1):158–65.

Torgen, M., L. Punnett, L. Alfredsson, and A. Kilbom. 1999. Physical capacity in relation to present and past physical load at work: A study of 484 men and women aged 41 to 58 years. *Am J Ind Med* 36(3):388–400.

Waters, T.R., V. Putz-Anderson, and A. Garg. 1994. *Applications Manual for the Revised NIOSH Lifting Equation*. Cincinnati: National Institute for Occupational Safety and Health.

de Zwart, B.C.H., M.H.W. Frings-Dresen, and F.J.H. van Dijk. 1995. Physical workload and the ageing worker: A review of the literature. *Int Arch Occup Environ Health* 68(1):1–12.

FURTHER READING

Åstrand P.-O., K. Rodahl, H.A. Dahl, and S.B. Stromme. 2003. *Textbook of Work Physiology. Physiological Bases of Exercise*. Windsor, Canada: Human Kinetics.

Hall S.J. 2006. *Basic Biomechanics*. 5th ed. Boston, MA: McGraw-Hill Higher Education.

Nordin M. and V.H. Frankel. 2001. *Basic Biomechanics of the Musculoskeletal System*. 3rd ed. Baltimore, MD: Lippincott Williams & Wilkins.

Wilmore, J.H., D.L. Costill, and W.L. Kenney. 2008. *Physiology of Sport and Exercise*. Champaign, IL: Human Kinetics.

4 Work in Awkward Postures

Karin Harms-Ringdahl

Photo: Karin Harms-Ringdahl

CONTENTS

Andrej, who is 32 years old, is 1.81 m tall and weighs 78 kg, has a live-in partner, and works as a painter. He works for a large company which is often commissioned to undertake new constructions. For Andrej it is mostly a matter of painting ceilings in flats. Some ceilings are spray-painted; others painted using a large roller on a long pole (Figure 4.1a). The ceilings are completed using a smaller roller or brush. This involves working with heavy equipment, which requires considerable effort from Andrej's arm and shoulder muscles, with his hands held at shoulder height or higher. In addition, he has to look up at the ceiling so as not to splash ceiling paint over the entire apartment (Figure 4.1b). The relatively heavy effort from his shoulder muscles, with his hands lifted up high at the same time as his neck is bent back, means that Andrej gets pains in his neck and shoulder muscles as well as headaches. This is exacerbated by time pressures in noisy new constructions, which makes it difficult to take regular breaks or to vary his work posture.

Andrej is touching up the paintwork in a kitchen, when the plumber arrives with a dishwasher on a trolley. This machine has to be connected up in a narrow space beneath a sink unit (Figure 4.2), as soon as the carpet fitter (Figure 4.3) has laid a protective floor covering. And then the electrician has to get in and connect the power, but first he has to access the ceiling light. The dishwasher has to wait. The narrow space under the sink unit means that they have to get down on their knees and twist their backs. At the same time, they carry out work with their hands, which makes demands on being able to see clearly what they are doing. The electrician, who is 47 years old, wears varifocal glasses, which means that he has to bend and twist his neck in order to focus at the correct distance, as varifocals have different focal lengths in the upper and lower part of the lens, and moreover have only a limited lateral field of view. If he tries to hold his neck in a more comfortable posture, his field of vision becomes blurred. Despite wearing knee-pads, his knees ache and the muscles in his back are sore.

FIGURE 4.1   As (a) and (b) painting involving one hand held high above the head. The neck is bent back so that the painter can see the results of his work. The paint roller, which has to be rolled evenly, requires great activation of the neck and shoulder muscles, compressing the joints of the neck in a position that is bent back. Photo: Christer Spångberg.

FIGURE 4.2 Installation of a dishwasher where the plumber is on his knees with one shoulder stretched forward in an extreme posture, attempting to push the machine into place. Photo: Karin Harms-Ringdahl.

4.1   FOCUS AND DELIMITATION

This chapter deals with what happens when workers need to carry out tasks in uncomfortable or *awkward* work postures, in which the joints come under strain close to the limit of their range of motion, something that happens when someone has to twist and bend in order to carry out their work. It is reasonable to assume that this is what we, in daily life, call *awkward work postures*, a concept that occurs in questionnaire-based surveys from the Swedish Work Environment Authority and in European statistics on work-related disorders. Along with keeping the back and joints in awkward work postures, sometimes workers are required to hold that position for

FIGURE 4.3 In carpet laying work and when welding carpet joins, the carpet fitter stands and crawls on hands and knees with his back bent and somewhat twisted and his neck slightly bent backward. Photo: Karin Harms-Ringdahl.

a fairly long time while carrying out work with their hands, often using a tool and with a demand for visual accuracy. Apart from uncomfortable work postures, this also means that the muscles are often working monotonously and repetitively for long periods (see also Chapter 5). At building sites, moreover, the work is sometimes carried out under hot or cold conditions (see also Chapter 9) and under conditions that imply that workers are dependent on one another's tasks, and therefore may experience stress resulting from waiting for someone else to finish (see also Chapter 7). This chapter, however, mostly deals with the strain put on the joints and back while working in twisted or stooped awkward work postures.

In this chapter you will find answers to questions such as:

- How are muscle activity and joint load affected, and thereby the risk of pain, when the back and neck are held in an uncomfortable, stooped posture while working?
- Why do people get knee joint disorders from squatting down for a long time?
- Why can someone experience the symptoms of a slipped disc if they put strain on their back while twisting or bending it at the same time?
- Why is it more strenuous to twist one's body into an extreme, rotated posture, so that it is possible to look backwards, despite which the biomechanically stressful moment does not increase in comparison with a more comfortable, neutral position?
- Why do the neck and shoulders ache when painting a ceiling for a long time?
- Why does the back ache when straightening up again after standing or sitting stooped forward?
- Why are people not particularly strong when working in positions close to the limit of their joints' range of motion?
- Can Andrej do anything at work to be able to perform his job tasks without all these problems?

4.2 PREVALENCE IN WORKING LIFE

Awkward work postures can be encountered, for example, in a number of industrial, agricultural, building, and patient care jobs. In the European Union (EU), ~45% (ranging between 24% and 66% among countries) of the working population report being subjected to tiring or painful postures at least a quarter or more of the time, including 15% who report being exposed all or almost all the time [Eurofound 2006].

In the building industry, tasks often result in work being carried out with the body in very awkward postures, as the building itself establishes the framework for the environment—which is not so easy to alter in relation to the task. While the building industry is dominated by male employees, the opposite is the case in the care sector. The demands of the job and the opportunity for influence are considered to be a factor that mediates work-related disorders [Karasek et al. 1981]. Table 4.1 shows that building workers nevertheless feel that they have more physically awkward work

TABLE 4.1
Proportion (%) of Workers between 1999–2003 Who Assessed Awkward Work Postures of Various Kinds and Opportunities of Influencing their Working Environment

Number of Employees 1999/2003	Percent		
Work Environment Issues:	**Men and Women All**	**Men All**	**Men Construction**
Experience at work			
– Has strenuous work postures	36	33	58
– Has strenuous heavy work	27	27	57
– Has strenuous, monotonous work movements	29	27	40
– The work is strenuous and inflexible	18	15	7
Several times every day compelled			
– To lift at least 15 kg each time	17	21	39
– To bend and twist in the same way several times an hour	26	24	31
Demands and influence			
– Cannot determine when different tasks are to be done	43	37	34

Source: Lundholm L. and Swartz H. 2006. *Musculoskeletal Ergonomics Statistics.* Report 2006:2E Swedish Work Environment Authority. http://www.av.se/dokument/statistik/english/Musculoskeletal_ergonomics_statistics.pdf.

Note: The number of men employed in the Swedish Construction Industry each year during the period was ~212,000.

postures, although they actually report that they have a less restrictive and a freer job than other men and women, respectively [Lundholm and Swartz 2006].

But computer work may also involve uncomfortable work postures, if the shoulder joints are kept in a (moderately) outward-rotated position for a long period, such as when using a mouse that is placed a little too far to the side of the keyboard [Karlqvist et al. 1998]. Computer work can also mean that the wrists are kept at a sharp angle for long periods of time, with entrapment disorders (the so-called entrapment of the median nerve in the carpal tunnel) as a result.

4.3 DESCRIPTION OF THE EXPOSURE

The design of the environment and those activities we carry out at work and in our leisure time, together with our body dimensions and movement habits, determine what positions put strain on our joints and how much of our relative muscle strength the task demands. Normally we try to work with our back, neck, and joints in relatively neutral positions (see Fact Box 4.1). Then the muscles help above all to stabilize the body and to ensure that the work can be done with the use of optimal force and movement. If the task allows, we alter our body posture so that the work can be carried out in the most

FACT BOX 4.1

The neutral zone in a joint—the area where it is *relatively* equidistant from the
limit of the range of motion in all directions. The ligaments that stabilize the
joint are relatively similar and equally lightly loaded. When the joint is kept in
a neutral posture, load occurs in the neutral zone of the joint. When the joint is
at the limit of its range of motion, this is termed as an *extreme posture* [Harms-
Ringdahl 1986; Harms-Ringdahl and Ekholm 1986].

comfortable way possible, which also means that we spontaneously vary our work
posture. When the task demands that manual work must be performed, and the worker
needs to control hand-held tools, workplace design, and the opportunities for its adaptation relative to the body dimensions of the employee have a decisive significance on
which work postures they adopt. Sometimes a conflict may also arise, which has to do
with the time factor. From a short-term perspective, it may take a longer time to optimize a work posture than to carry out the task in a more awkward body posture, which
may not hurt at the time. The use of glasses may sometimes also be a factor, which
means that a worker has to bend and twist the cervical spine to be able to see properly.
A common problem for people who have acquired glasses as a result of age-related
changes in their eyesight (difficulties of focusing at short distances) is the difficulty of
gaining visual acuity for different distances. Varifocal glasses mean that the size of
the optimal field of vision and visual acuity can be adjusted by changing the head position and thereby the angle of the neck. Someone without glasses can see out of the
corner of their eye or in the upper or lower part of their field of vision, and in this way
one is able to keep the cervical spine in a more comfortable, neutral position.

In Andrej's case, there is a given height to the ceiling, and the ceiling has a given
area that has to be painted. This means that he needs to bend his neck back to be able
to see upto the ceiling while he is working. A brush or roller is certainly light, but
once dipped in paint, its weight increases considerably. In addition, he needs to press
the roller or brush against the surface and lift his arms so that his hand is raised
above the head height in order to be able to reach. If Andrej uses a roller on a long
pole (Figure 4.1a), he can hold his hands lower than when he is painting the ceiling
with a shorter roller. As a roller on a long pole can be held in front of his body, he
does not need to bend his neck as far back in order to see as and when he is using a
roller with a short pole (Figure 4.1b). At the same time, a long pole means that considerable force is required for the muscles of the neck and shoulders to manoeuvre
the tool in an efficient way against the surface of the ceiling.

Dishwashers have to be connected with pipes and electrical installations, located
far inside cupboards under sink units, which means that the installer has to twist in
order to reach (Figure 4.2). Often he has to work on his knees to get into this kind of
narrow space.

Electrical work means drawing cables across the ceiling using great precision and
sometimes a heavy hand-held tool, and while standing on a ladder. Using a more
uncomfortable work posture is easy to choose over taking the time to move the
ladder again and again so as to reach more comfortably. The requirements for precision

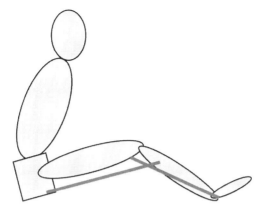

FIGURE 4.4 Illustration of how muscles that pass two joints (hip and knee, respectively, knee and wrist) act together to provide a stooped position in the lumbar spine and cervical spine. The Achilles tendon attaches to the back of the heel and forms a lever vis-à-vis the rotation centre of the foot joint. Illustration: Christer Spångberg.

work and the importance of the hand-held tool, as well as the hand and arm strength needed to carry out the task, add to the strain caused by the body posture itself.

The strain at a joint's extreme posture (see Fact Box 4.1) may cause a temporarily limited range of motion in another adjacent joint. This is the result of some muscles, for example, those at the back of the thigh, passing through two joints. An example of an effect of this kind is when driving a car with the driving seat in a low, pushed-back position (Figure 4.4). The position of the accelerator pedal means that the driver's right leg is stretched forward so that the foot can reach the pedal, which means that the knee joint is almost straight. The narrower the angle between the thigh and the trunk, the more the muscles at the back of the thigh tighten when the knee joint is held outstretched. It is then difficult to sit and at the same time retain curvature in the lumbar spine. The driver has to sink with a rounded lumbar spine into a completely bent position, which at the same time means that the cervical spine is bent back further and the chin is pointed so as to be able to keep one's eyes on the road. At the same time, the opportunity of twisting the cervical spine is limited, and it becomes more difficult to turn to look back over one's shoulder into the "dead angle."

4.4   NORMAL PHYSIOLOGICAL RESPONSES AND MECHANISMS

The joints in the body are a precondition for mobility. The joint is a functional unit consisting of joint surfaces, ligaments, and joint capsule, which are covered on the inside by the synovial membrane, which produces synovial fluid, which in turn increases the capacity for gliding. The joint surfaces, which have various anatomical designs, are constructed so as to create good contact between the surfaces, and are usually more or less arched. The joint surfaces are covered in cartilage, which makes it possible for them to glide against each other with very little friction. In movement, a translatory sliding movement in the plane of the joint surfaces and an angular,

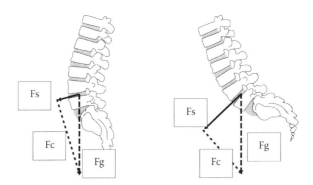

FIGURE 4.5 Work in a position that is markedly stooped puts strain on ligaments and connective tissue around the vertebrae. The illustration shows the bodies of the vertebrae in the lower lumbar spine with the intervening discs seen from the side (on the left in a neutral, upright position; on the right in a position leaning forward). Fg—gravitational force, Fc—compressive force, and Fs—translational force (i.e., shear force). Illustration: Niklas Hofvander. Modified after Christer Spångberg.

biased rotation movement take place. Strain on the joint creates both compressive forces perpendicular to the joint surface and translatory forces, depending on the directional force of the load (Figure 4.5).

Sometimes there are cartilage discs, the so-called meniscuses as in the knee joints (Figure 4.6), and vertebral discs as in the back (Figure 4.5) and neck, which help distribute the compressive force over a larger area, ensuring that the load per unit area is reduced.

Joint cartilage has no blood vessels and receives its nutriment through osmosis—that is, an equalization of concentration through a membrane; it benefits from var-

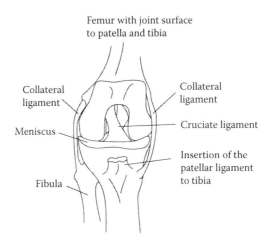

FIGURE 4.6 Knee joint (without the kneecap) seen from the front with the most important stabilizing ligaments—cruciate ligaments and collateral ligaments. Illustration: Niklas Hofvander. Modified after Christer Spångberg.

ied load, for example, as happens in the cartilage of the hip when walking. In varied load, the cartilage is compressed and the load then released, which provides an effect as though squeezing out a bath sponge. This increases the supply of nutriment to the joint cartilage (see also Chapter 3, Section 3.8.2). As most of our movements in daily life occur at the neutral position of the joints, the joint cartilage is, relatively speaking, thicker in the central areas of the joint surfaces—the neutral zone of the joint—while it is thinner in the more peripheral, outer parts, and is in those parts more sensitive to load.

The joints are stabilized by ligaments and joint capsules, which centre on the load and prevent the joint from luxating ("dislocating"). The ligaments perform an important function in ensuring that the joint is stable at rest and in movement (Figure 4.6). When the joint is under strain at the limit of its range of motion, some of the ligaments are greatly tensed and the joint surfaces are pressed together in the extreme position of the joint, which impairs the exchange of nutriment to the cartilage and accelerates wear. This happens, for example, in the knee joints when squatting or kneeling. Prolonged uneven strain can lead to wear, in that the nutriment exchange to the cartilage becomes impaired.

Sudden overloads can also damage the various structures of the joint. This might, for example, be a question of a healthcare worker who catches a falling patient, which may lead to powerful compressive force in the worker's spinal column, which among other things compresses the vertebral discs of the lumbar spine. A compression of this kind can affect the vertebral disc so that its viscous content is pressed out, leading to what is known as a slipped disc. If a slipped disc presses on the nerves emanating from the spinal column, painful symptoms appear. As these nerves in the lumbar spine reach down primarily to the legs, the pain or numbness that then occurs will be perceived as coming from the affected leg. Pain of this kind has been called after the nerve which goes down into the legs—sciatic pains. Putting great strain on the spine through heavy lifting, for example, particularly with a twisted or flexed back, also results in an uneven strain on the vertebral discs, which further increases the risk of a slipped disc.

Connective tissue around muscles and in tendons and ligaments has viscoelastic properties, which means that it has a length adapted to the physiological range of motion and which, with varied everyday loads, returns to its original length. The physiological range of motion normally varies somewhat between individuals; however, some people regard themselves as "stiff" while others see themselves as "very supple." If connective tissue is stretched for a long period with a particular force, the tissue lengthens and its tension declines—known as "creeping phenomenon"—and the range of motion increases even after the stretching force is removed. If, instead, one avoids making use of the normal, physiological range of motion in a joint, for example, when an arm or leg is in plaster or when it is painful, the connective tissue is shortened and the range of motion is reduced. Varied moderate load in the various positions of the joint on the other hand instead helps build up the cartilage, and the joint becomes more resistant to load. Wear on the connective tissue may mean that the original length is altered, and thereby the biomechanical conditions in the joint. If the ligaments are damaged and held in an extended position, the stability of the joint is affected and there is a risk of wear on the cartilage of the joint surfaces, known as arthritis, which in turn makes the joint more sensitive to load than normal (see also Chapter 3, Section 3.8.2).

The joints are surrounded by muscles, both shorter muscles which only serve one joint and longer ones serving two or more joints. The muscles attach to the bone with a structure of tendons. The surrounding muscles are important both for the stability of the joint and its ability to cope with loads, as well as counteracting external loads and achieving movement.

When joints are under strain near the limit of their range of motion, the ligaments are stretched as well as other connective tissue structures surrounding the muscles and serving the relevant joint. Stretching the muscles and tendons in extreme postures activates the Golgi tendon organ, which signals that the tendon is being stretched. After a while, depending on load and time, the pain receptors are stimulated and we feel pain, a pain, the intensity of which instantaneously increases at the moment when returning to the starting point, and which can persist for a long time even though we have started to move again.

The muscles' ability to develop force is also dependent on the relative muscle length at a certain angle, and, moreover, co-varies with the tendon's moment arm (lever) to the joint's axis of rotation (Figure 4.7). The strength (force × moment arm) one can develop at a certain joint angle varies at different angles of the joint. Given the same moment arm length a somewhat extended muscle can produce more force than a shortened, but the overall strength of the muscle anyhow declines the nearer one comes to the extreme joint position, due to shorter lever arms. When the joint is kept in its extreme posture, those muscle groups that counteract the load—that is, those that can bring the joint back to a more neutral position—are very extended and their moment arm shortened. The strength the individual can develop in this situation is therefore more limited. As working life often demands that a certain operation is carried out with a certain force, a relatively greater proportion of the strength capacity is used to carry out that operation compared with carrying it out with the joints and muscles in a more comfortable work posture (see also Chapter 3).

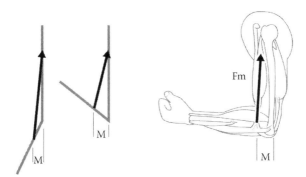

FIGURE 4.7 Strength in the elbow flexors (Fm × M) depends on muscle force Fm in the biceps muscle and the tendon's moment arm (the perpendicular distance M) to the elbow joint's axis of rotation. On the left a schematic image showing examples of two different elbow angles with varying muscle length and moment arm. Illustration: Christer Spångberg. Adapted by Niklas Hofvander.

FIGURE 4.8   Work in a markedly stooped posture puts strain on ligaments and connective tissue around the vertebrae. Photo: Christer Spångberg.

On the other hand, help can be received from the stretched connective tissue structures to counteract the strain in an extreme posture. An individual with, relatively speaking, less muscle strength can then exploit the connective tissue structures to counteract the strain, and does not then need to develop any muscle activity. Gardening, which means that one needs to flex the lumbar spine to the limit of its range of motion, is one example of an awkward work posture in which the level of activity in the back muscles is low while the mechanical load on the structures of the back is high (Figure 4.8). The load comprised by body weight is carried completely by the ligaments around the vertebrae and the fasciae of the connective tissue around the muscles. The muscles in the back of the leg are also extended considerably, but there the activity levels in the muscles help keep the body in balance.

The increased strain on the joints in extreme posture contributes to an increase in both the compressive forces and translatory forces in the back (Figure 4.5). The construction of the back is optimized so as to take up compressive forces, while its tolerance to translatory forces, which increase the strain on the joint capsules and ligaments, is less sound.

While the muscle activity subsides, despite the fact that the load moment increases when bending one's back and putting load on the joints in their extreme posture with the aid of gravity (see Figure 4.8), the reverse is the case when twisting one's body around its own longitudinal axis. In twisted body postures, the muscle activity increases logarithmically with the angle of rotation, without the load moment itself increasing, the closer to the limit of the range of motion one comes, as the ligaments and joint capsules are tightened. In that situation, a mechanical load in an awkward posture is combined with a high muscular static load to keep the joint in its extreme posture [Torén and Öberg 1999].

A loading moment can therefore be counteracted both by muscle activity and/or connective tissue structures and trigger pain depending on intensity and duration [Harms-Ringdahl 1986; Harms-Ringdahl and Ekholm 1986].

4.5   POTENTIAL NEGATIVE RESULTS OF EXPOSURE

4.5.1   Does it Hurt?

We have all walked bent over for a long time picking berries, weeding the garden or slumped down in a sofa, and have then wondered why it is so painful in the lower back when we get up. The pain lasts for a while, but can also reappear as a kind of delayed onset muscle soreness, like the day after physical exercise, depending on how long the extreme posture load has lasted. The slow and dull ache points to the fact that the slower, inward-reaching pain nerve fibres modulate the pain.

Why a prolonged, but, relatively speaking, light stretching at the extreme posture of the joint should lead to tenderness and pain when moving in otherwise entirely pain-free individuals is not completely known, but there are several theories.

There are many nociceptive receptors (pain-sensitive nerve fibres) in the connective tissue structures, as well as ligaments, joint capsules, connective tissue fasciae, and tendons, which can be stimulated *mechanically*. The cartilage structures themselves, on the other hand, have no pain receptors. But if the cartilage in a joint is damaged, this can give rise to a painful reaction in the joint. A change in the *chemical* environment of the nociceptive receptors can also stimulate the pain receptors in the ligaments and periosteum [Wenngren et al. 1998]. There is even speculation that the raised pressure occurring inside the joint and in the bone tissue itself, when it is under strain near the limit of its range of motion, might produce pain.

Pain as a result of extreme posture can also generate a continuous muscle reflex activity which can contribute to activating muscle spindles and substances that trigger pain in the tissues [Johansson et al. 1999].

Within veterinary medicine, provocation load in extreme postures is used to chart lameness in horses as a result of joint disorders; the joints in one leg at a time are flexed at the same time as some of the weight of the horse rests on the flexed joint for a standard number of seconds. Then the load is released and the horse is allowed to jog for a few paces to see whether lameness is present. A corresponding argument can be adduced for human beings. If, for any reason, someone has pain in a joint, the pain-free range of motion is restricted. The extreme posture approaches the neutral zone and it is painful when they use their own muscle force or when a clinician flexes the joints to the limit of the now restricted range of motion.

4.6   INCIDENCE OF DISORDERS

Based on EU27 statistics from 2005, in which 235 million people from 31 countries participated, 45% of EU workers in the construction sector and 39% in the health sector reported that their health had been affected by their work; backache and muscular pain were the most frequently reported physical symptoms [Eurostat 2009].

The risk of developing a disorder as a result of awkward work postures depends, to a great degree, on the size of the workload in the extreme posture and the time the exposure lasts. This means that it is difficult to assess the actual incidence of hazardous exposure to awkward work postures in working life. According to EU statistics from 2005, 25–66% of the workers (a mean of 45%) in 31 countries report exposure to tiring or painful positions at least a quarter of the time, of whom 18% report such exposure all or almost all the time. The figures have remained relatively similar over the last 10 years [Eurofound 2006].

Awkward work postures are the most frequent cause of disorders among men and the second most frequent cause among women. One in five male painters, chimney sweeps (23.2%), and building and construction workers (21.0%), one in six motor vehicle mechanics and motor vehicle repairers (18.3%) and electrical installers (18.1%) in Sweden report disorders as a result of awkward work postures [Weiner and Bastin 2005].

The equivalent applies to one in four female hairdressers and skin therapists (27.0%) and one in five cooks (23.3%), assistant nurses (20.3%), employees in institutional households and restaurants (19.2%), and hotel cleaners (19.1%).

Why a somewhat higher proportion of women than men report disorders as a result of awkward work postures is unclear, but there may be many reasons, such as women are more likely to report disorders or that tools and work heights often are better adapted to men's hand and body sizes, which makes the work postures more stressful for a smaller person. Direct measurements show, however, that women have more awkward work postures than men. This applies primarily to those who have little influence over their work situation and who work in professions where women are predominant [Leijon et al. 2005].

Among women-dominated professions under county council, municipal and private direction, there is a 60% greater incidence of reported work-related disorders among dental nurses (after adjusting for the difference in age distribution) (period prevalence = 1.6) [Weiner 2006]. The tasks of dental nurses involve both awkward work postures and prolonged static muscle work in combination (see Chapter 6).

4.7   RISK ASSESSMENT

There are different methods of quantifying the incidence of awkward work movements. Observations over a working day or several days are time consuming, whether these observations are carried out by an on-site observer who then enters all the values manually into a computerized system, or are done using video-recorded data. Problems occur, however, when the work is mobile and cannot be reproduced using two-dimensional observations. A more objective method is the use of an angular gauge that can be applied to the joints in question and which provides angular values around three axes and computerized signals. Here the work movements can be related to percentage proportions of the range of motion of the relevant joint around the corresponding axes.

In an experimental study, subjects have been asked to assess how comfortable/ uncomfortable it is to keep their shoulder joint and neck in a given position in different parts of its range of motion [Kee and Karwowski 2001]. Even though this was done for a very short time and with individuals who had no disorders, for example, it was felt to be uncomfortable even after 60 s to hold one's arm rotated outward 25° from the shoulder. It is common for the computer mouse to be placed next to the keyboard, which causes this precise outward-rotated shoulder position [Karlqvist et al. 1998]. As regards the neck, all directions of movement were regarded as increasingly uncomfortable the further one deviated from holding one's head in a neutral posture [Kee and Karowski 2001].

Assessments of the degree of exertion, and discomfort or pain are also frequently carried out with the aid of various types of scale. The more specific the question asked, the more specific was the response received. In statistical studies of the working environment in the EU (e.g., the Fourth European Working Conditions Survey), questions are asked, for example, about whether one experiences work postures as painful or tiring and awkward. One problem with questions of this kind is that it is not possible to distinguish whether it is the tasks in awkward positions or the work of muscles in various work postures which mean that the posture is experienced as strenuous or awkward. People are, however, as a rule, good at making various types of assessments of experiences with a high degree of reproducibility, assuming that they are asked a precise question [Leijon et al. 2002]. On the other hand, it is more uncertain as to how possible disorders influence the assessment of exposure and its duration. It is more probable that a person with a disorder in a particular work posture will report it as awkward and of longer duration compared with a person without any disorder in the same work posture. It is also very possible for someone to change their work postures and try to avoid awkward positions as far as possible if they have disorders triggered by strain.

Another possibility in risk assessment is that people who are sick and off work for a long period or who have changed jobs or ended their employment as a result of any disorders are not picked up in the study. Studies in which a group of individuals are followed over time are therefore to be recommended.

4.8 MEASURES IN CRITICAL CONDITIONS

The most important measure is to revise work postures and tasks so that load on the body and work movements varies naturally during the working day. Moderate load in varied work postures reduces the risk of disorders, while inflexible work postures increases the risk, irrespective of the external load and position of joints.

While seated work sites can often be altered to make it easier for the employee, there are many examples of environmental factors that cannot easily be adjusted, and where instead it is necessary to consider whether there are any work aids to reduce the strain on the body and in this way facilitate variation and the carrying out of tasks (Figure 4.9). In the left-hand-side picture, one can work with one's back and joints in comfortable work postures. In the right-hand portion of the picture the work is not altogether made easier by using a step. As the task is so close to ground level, both back and neck are kept in awkward work postures. In this case, the entire vehicle should instead have been raised with the help of a lift to facilitate the work. Presumably, it would also have been better to carry out the work standing up.

FIGURE 4.9 A simple moveable step (a) allows work to be carried out without needing to twist or stoop in order to gain access. On the other hand, working on a step—depending on the work height—may also mean that it is more difficult to gain access (b). Photo (a): Christer Spångberg; (b): Karin Harms-Ringdahl.

Preventive measures can be divided into two groups—Technical devices and Organization of work. While different types of aid can adjust the work height for the employee, so that the work postures become comfortable and inflexible body postures are avoided, work organization measures can build in natural variation. Different measures by which the employee is reminded at regular intervals to take breaks without these occurring naturally have most often not resulted in the intended effect.

Those aids available on the market often provide opportunities for changing the work height–employee relationship. Different types of simple steps are an example of this (Figure 4.9). Other examples may be lengthening the pole on a paint roller, or adapting the tool so that it is easier to reach, and creating good lighting at the workplace. In work in restricted spaces, it is often easier to see what should be done in a strong light. For our spectacle-wearing electrician, the work area which he can see close up is much larger in a good light. There is also the possibility of using bifocal glasses with the lenses reversed, showing the immediate surroundings at the top of the lens, which reduces the need to bend the neck back in carrying out precision work at above head height.

From a work organization point of view, there is an attempt to create the preconditions for body movements being variable both as regards range of motion and muscles. Changing ingrained movement habits, however, takes time and requires practice. On Andrej's part, one might imagine that his work could be reorganized so that he does not have the same task all day, or perhaps even for several days in a row. Vertical work could be alternated with tasks in the lower area of movement.

Training forms part of the work organization measures, even if the results of providing information about ergonomics are not unequivocal. This might be because it is merely information, and because a worker is seldom given the opportunity of training the performance of work movements in the same systematic way that a sportsman, for example, a golfer, trains the different movements needed to pursue his sport in a way that is optimal for his body.

4.9 WHAT DOES THE LAW SAY ABOUT WORK IN AWKWARD WORK POSTURES?

For the EU member states, there are a number of minimum directives applying to the work environment and health [Directive 89/391/EEC, Directive 89/654/EEC, Directive 89/655/EEC]. There are no detailed rules in these directives applying to awkward body postures, but it is clearly evident that employers have a responsibility to perform a risk assessment on all jobs as regards ergonomics and to take measures above all to remove risks, and secondly to minimize them.

Another piece of European legislation that has a bearing on work postures is the directive regulating the use of personal protective equipment [Directive 89/656/EEC]. In this directive there are demands that personal protective equipment should be designed ergonomically and adapted to the employee. It is important to reduce the physical strain in jobs involving awkward body postures at the same time as personal protective equipment is being used.

The member states have national legislation to implement and concretize the directives. There are often national guidelines and general advice as a support for measures to prevent risks at work.

Regulations about work postures that should be avoided are given in very general terms and as a rule issue from the muscle load and body postures which may be thought to cause disorders, how heavy a weight one should lift, push or pull, and how still one should sit. For awkward work postures, in principle the corresponding regulations for sitting apply (see further in Chapter 6) as for standing and walking tasks. Work using the back and neck in stooped and/or twisted work postures, like work with the hands above shoulder height or below knee height, is regarded as a risk factor for disorders triggered by strain, if this occurs periodically or over a large part of the working period.

There are checklists [Kemmlert 1995] and models for identifying and assessing awkward work postures, where risk factors for disorders triggered by strain are evaluated in three risk factors: red, yellow, and green. Organizational measures for achieving variation of work movements should be aimed at, while awkward work postures, such as prolonged stooped or twisted positions with high demands on vision, maintained periodically or for long periods, should be avoided.

4.10 SUMMARY

Work with a high demand for work postures in which the extremity joints and the back and neck are under strain close to the limit of range of motion is normally quite frequent, perhaps particularly in the building trade but also in other types of work. Approximately 8–10% of the population state that they have disorders of the musculoskeletal system as a result of awkward work postures. Loads close to the limit of the range of motion are counteracted by connective tissue structures, and may be associated with little or a great deal of muscle activity, depending on whether the position can be maintained with the help of gravity or whether it is necessary to twist around a longitudinal axis. The cartilage is thinner at the limit of the range of motion and more sensitive to major compressive forces than if the corresponding force impacts on the cartilage in the central portions of the joint. We are also less strong in the angles of the joint near the limit of the joint's range of motion. Forces parallel to the surface of the joint (translatory forces) that stretch the ligament and joint capsules can arise. Pain occurs after a period of strain in an extreme posture, and more rapidly if the individual already has a disorder in the joint concerned. There are pain-sensitive nerve fibres in the connective tissue structures that help generate pain. Countermeasures consist of ergonomic changes, and changes to the organization of work, that promote varied body movements and load in more neutral joint positions.

REFERENCES

Eurofound. 2006. *Fourth European working conditions survey.* European Foundation for the Improvement of Living and Working Conditions. http://www.eurofound.europa.eu/pub-docs/2006/98/en/2/ef0698en.pdf

Eurostat. 2009. *Statistics in focus.* 63/2009. http://epp.eurostat.ec.europa.eu/cache/ITY_OFFPUB/KS-SF-09-063/EN/KS-SF-09-063-EN.PDF

Directive 89/391/EEC—*On the introduction of measures to encourage improvements in the safety and health of workers at work.* European Agency for Safety and Health at Work. http://osha.europa.eu/en/legislation/directives/the-osh-framework-directive/1

Directive 89/654/EEC—*Concerning the minimum safety and health requirements for the workplace.* European Agency for Safety and Health at Work. http://osha.europa.eu/sv/legislation/directives/workplaces-equipment-signs-personal-protective-equipment/osh-directives/2

Directive 89/655/EEC—*Concerning the minimum safety and health requirements for the use of work equipment by workers at work.* European Agency for Safety and Health at Work. http://osha.europa.eu/en/legislation/directives/workplaces-equipment-signs-personal-protective-equipment/osh-directives/3

Directive 89/656/EEC—*On the minimum health and safety requirements for the use by workers of personal protective equipment at the workplace.* European Agency for Safety and Health at Work. http://osha.europa.eu/en/legislation/directives/workplaces-equipment-signs-personal-protective-equipment/osh-directives/4

Harms-Ringdahl K. 1986. On assessment of shoulder exercise and load-elicited pain in the cervical spine. Biomechanical analysis of load-EMG-methodological studies of pain provoked by extreme position. *Scand J Rehab Med Suppl* 14:1–40.

Harms-Ringdahl K. and Ekholm J. 1986. Intensity and character of pain and muscular activity levels elicited by maintained extreme flexion position of the lower cervical-upper thoracic spine. *Scand J Rehab Med* 18:117–126.

Johansson H., Sjölander P., Djupsjöbacka M., Bergenheim M., Pedersen J. 1999. Pathophysiological mechanisms behind work-related muscle pain syndromes. *Am J Ind Med Suppl* 1:104–106.

Karasek R., Baker D., Marxer F., Ahlbom A., Theorell T. 1981. Job decision latitude, job demands, and cardiovascular disease: A prospective study of Swedish men. *Am J Public Health* 71:694–705.

Karlqvist L., Bernmark E., Ekenvall L., Hagberg M., Isaksson A., Rosto T. 1998. Computer mouse position as a determinant of posture, muscular load and perceived exertion. *Scand J Work, Environ & Health* 24:62–73.

Kee D. and Karwowski W. 2001. The boundaries for joint angles of discomfort for sitting and standing males based on perceived discomfort of static joint postures. *Ergonomics* 44:614–648.

Kemmlert K. A method assigned for the identification of ergonomics hazards—Plibel. 1995. *Appl Ergon* 26:199–211.

Leijon O., Wiktorin C., Härenstam A., Karlqvist L., MOA Research Group. 2002. Validity of a self-administered questionnaire for assessing physical workloads in a general population. *J Occup Environ Med* 44:724–735.

Leijon O., Bernmark E., Karlqvist L., Härenstam A. 2005. Awkward work postures: Association with occupational gender segregation. *Am J Ind Med* 47:381–393.

Lundholm L. and Swartz H. 2006. *Musculoskeletal Ergonomics Statistics*. Report 2006:2E Swedish Work Environment Authority. http://www.av.se/dokument/statistik/english/Musculoskeletal_ergonomics_statistics.pdf

Torén A. and Öberg K. 1999. Maximum isometric trunk muscle strength and activity at trunk axial rotation during sitting. *Appl Ergon* 30:515–525.

Weiner J. 2006. *Arbetsorsakade besvär i landsting och privat sektor—en jämförelse*. Stockholm: Swedish Work Environment Authority (In Swedish). http://www.av.se/dokument/statistik/sf/sf2006_02.pdf

Weiner J. and Bastin M. 2005. *Work-Related Disorders* 2005. Stokholm: Swedish Work Environment Authority and Statistics Sweden. http://www.av.se/dokument/statistik/officiell_stat/ARBORS2005.pdf

Wenngren B. I., Pedersen J., Sjölander P., Bergenheim M., Johansson H. 1998. Bradykinin and muscle stretch alter contralateral cat neck muscle spindle output. *Neurosci Res* 32:119–129.

FURTHER READING

Rom W. N. and Markowitz S. B. 2007. *Environmental and Occupational Medicine*. Philadelphia: Lippincott Williams and Wilkins.

5 Work with Highly Repetitive Movements

Fredrik Hellström

Photo: Johan Karltun

CONTENTS

Janis, 27, has worked for 5 years as a meat-dresser in a large slaughterhouse. As a meat-dresser, Janis dresses the carcasses of pigs that have been rough-butchered by a butcher. The carcasses are divided into halves or quarters that are either lying flat or hanging on a hook. The work involves separating the meat from the bone; portions are cut out and the fat is removed. The cuts of meat that are divided are placed on a conveyor belt, which takes them to sorting and packaging. Janis carries out most of the operations manually and works together with fellow workers in a team. The incidence of rotation between different tasks is limited. The different operations that Janis carries out make varying demands on strength and precision and take between 40 and 60 s to complete. Janis often works on one operation for a whole day, for example, cutting shoulder joints from pigs. In his work, Janis lifts weights of up to 15 kg; occasionally these peak as high as 25 kg. During his work, the force he needs to use to cut varies between 11% and 30% of his maximum force, with peaks at over 50%. The pace of work is high, and the time allocated for breaks is short. Janis begins at 6:00 in the morning and finishes work at 2:30 in the afternoon. He takes a total of 30 min break, divided into two periods during the day. His wages are in part performance-related, and so Janis wants to do as much work as possible during the working day. It is cold in the premises where Janis works. To protect himself against cuts, he has to wear protective gloves and a protective apron. Janis is often physically exhausted at the end of the working day.

5.1   FOCUS AND DELIMITATION

This chapter deals with work in which movements are repeated over and over again, at the same time as requirements are imposed for precision in movement and the muscles are used with considerable force. Janis' job as a meat-dresser in a slaughterhouse is a good example of this type of work. The chapter also illustrates how repeated movements, and the requirements for precision and force, are combined during the entire day or parts of the working day.

The focus in the chapter is on short repeated movements, and the chapter therefore deals only with movements of a maximum duration of five min. It is common, however, that movements are much shorter (between 5 s and 2 min). There are also jobs that entail large elements of repeated movements and precision demands, but which do not require the muscles to be used forcefully, for example, computer work. Computer work will be discussed in Chapter 6. Jobs that do not have a very repetitive element, but where considerable force is needed—called heavy manual trades— have been discussed in Chapter 3.

This chapter focuses on the physiology behind the performance of highly repetitive movements, and how highly repetitive movements with demands for force and precision may lead to disorders. The chapter will, among the things, answer questions such as:

- What preconditions does the body have for carrying out repetitive work?
- Why is a prolonged exposure to the combination of repetitive movements, force, and precision particularly taxing?
- Is working technique in repetitive movements of any significance?
- What happens to the muscles in repetitive work?

5.2   PREVALENCE OF HIGHLY REPETITIVE WORK IN WORKING LIFE

Janis is not alone in being subjected to highly repetitive movements, and force and precision demands, in his work. Various professions where highly repetitive work is common are to be found primarily within difference sectors of the food industry, for example, in the slaughterhouse sector, including the preparation of meat products, or the fish processing sector, and within the manufacturing industry, focusing on assembly work, particularly the assembly of motor vehicles. In Europe ~1 million people worked as butchers, meat-dressers, head butchers, or pork butchers in 2006, with most of these workers in countries such as Germany (202,000), France (157,000), and Poland (125,000). Approximately 130,000 people worked in the fish processing industry, in Europe, with predominance in countries such as Spain, the United Kingdom, and France. During 2006, ~2.2 million people worked on motor vehicle assembly in Europe, with predominance in Germany (980,000), France (416,000), and the United Kingdom (326,000). During the financial crisis of 2008–2009 the motor vehicle industry was hard hit, which means that the figures may have changed. A large proportion of these people worked in systems with line production, the so-called assembly lines, with either a continuous movement of the line or with a system where the line stops for a limited time while the task is performed. The system of assembly line production means that the work is more constrained and has a greater element of repetitive operations than in systems with greater autonomy as regards movements and tempo.

In total, in Europe, 62% of the working population state that they have been exposed to repetitive hand or arm movements for at least a quarter of their working day [Eurofond 2007]. The statistics collated from different member states in the European Union (EU) on exposure to repetitive work are not entirely uniform, but may also include exposure to considerable demands for muscle force and precision, or less strenuous jobs with repetitive elements. The statistics are therefore not completely transferable to those professions mentioned above, but nevertheless give a picture of the situation.

5.3 REPETITIVENESS AND EXPOSURE

5.3.1 THE WORK CYCLE AND ITS ELEMENTS

To perform an operation and return to the initial position again is called the work cycle, and the time it takes is called cycle time. Usually, the work cycle contains one or more operations that are repeated with high frequency. In Janis' case, it takes about a minute to cut out a whole joint of meat, which may be regarded as one work cycle, but during that minute Janis carries out a number of cutting movements with his arm and hand. The cutting movements consist in their turn of very short repeated movements (1–2 s of cycle time). A single repeated movement that forms part of a work cycle might be called a cycle of movements. The cycle of movements also has a cycle time. A work cycle may therefore consist of differing levels, where shorter operations with similar movements—cycles of movements—build up into the more comprehensive work cycle.

5.3.2 WHAT DOES EXPOSURE LOOK LIKE?

When Janis is working, it takes a minute to cut out a shoulder. During that minute he executes ~30 cutting movements with his knife. If Janis works cutting out shoulders from 6.00 in the morning to 2.30 in the afternoon, he makes at least 12,000 cutting movements with his arm and hand during his working day. In order to be able to cut, Janis needs to use ~20% of the maximum strength (maximum voluntary contraction (MVC)) in his arm, at the same time as it is important that he cuts accurately, as the product has to maintain a high quality. On a few occasions Janis may need to use more than 50% of his MVC in his work. The maximum force that Janis can generate in a particular movement is, however, dependent on the speed with which the movement is made. The quicker the movement, the lower the maximum force the muscle can generate. When Janis cuts out shoulders during the work cycle, he has a maximum speed in his shoulder movements of up to 315°/s and an average speed of ~65°/s. This means that it is probable that Janis is close to the maximum of what can be achieved in this specific movement.

He also needs to hold a knife, and for this he needs to use ~30% of his maximum grip force on average. It is common that the force required varies a great deal over the work cycle. The variation is illustrated in Figure 5.1 where the percentage of time of different grip force levels vary during different time periods of a work cycle when a butcher is preparing a piece of prime rib of beef. A grip force of 50–60 N corresponds to 20–40% of the maximum grip force, which is relatively powerful.

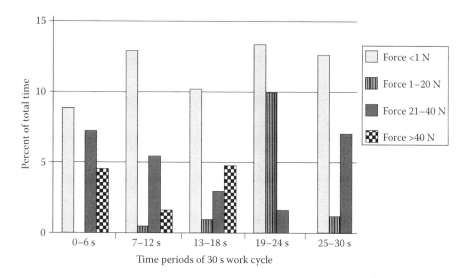

FIGURE 5.1 Different grip forces during different time periods in one individual cutting a piece of prime rib of beef. The percent of total time at different force levels are shown in each time period of the 30 s work cycle. N = Newton. (Modified from McGorry, RW., Dowd, PC., Dempsey, PG. 2003. *Appl Ergon* 34:375–82.)

In order to access the right places, Janis needs to position his body appropriately. This sometimes means stooped and twisted body postures along with keeping his arms above shoulder height. More about working in extreme body postures can be found in Chapter 4. At the same time, as the body posture is important, the position of the head must be carefully monitored, as eyesight is important when all the cuts must be made in the right places during dressing. This means that the muscles of the neck need to be tense all the time to keep the head still. The requirements on cutting in a particular way restrict the opportunities for otherwise carrying out the work in different ways; the work posture becomes fixed.

The neck muscles also help to stabilize the shoulders. As precision is higher in the hand than in the shoulder, the shoulder is used to roughly adjust the position and subsequently stabilize the arm. This puts continuous strain on the neck muscles.

The material that Janis is working with is also of importance to the exposure; if the meat is, for example, too cold, it becomes more laborious to cut into it, because cold meat is stiffer and harder. The exposure therefore consists of a combination of continuous and dynamic loads with considerable force development, carried out in a relatively fixed work posture (see Chapter 6 about dynamic muscle work). The same combination is also well documented in the fish processing industry and the assembly industry. When boning fish, activity of up to 56% of maximum electromyography activity has been measured in the muscles of the forearm, and a grip force of 31–34% of maximum hand grip force. Cycle times for different operations in the fishing industry vary between 5 s and 60 s, where operations of 5 s cycle time involve handling objects with weights of up to 2 kg. Operations with cycle times of 60 s involve handling objects of up to 21 kg [Nordander et al. 1999]. The work described above is also combined with a fixed work posture, primarily for the neck.

5.3.3 VARIABILITY OF EXPOSURE

Even if the task is the same, the relative load which muscles and joints are exposed to varies between individuals, as all individuals do not, for example, use the same proportion of their maximum force. For example, the hand grip force varies in a particular operation in the butchering of a pig between 11% and 35% of MVC, when different butchers are compared. The same difference in force between individuals has also been shown in assembly work using laminate sheets (installation, gluing, and final polish). This distribution of the force employed between individuals may depend on several different factors. Maximum strength (MVC) may vary among individuals, and/or individuals solve the problem in different ways. In order to be able to solve the task in different ways, it is necessary that the task allows this. By measuring the difference in how different individuals carry out the task, we can acquire some understanding of how inflexible the task is. The more flexible the task is, the greater the opportunity for differences between individuals. It is, however, not certain that everyone exploits or is even able to exploit the opportunity that exists to vary their work within those parameters that the task allows. Variability within the individual's performance of a task compared with that of other individuals may provide information about how similar the exposure is between work cycles, and thereby the flexibility in that individual's working technique.

5.3.4 REPETITIVENESS AND SIMILARITY

Repetitive movements are movements carried out in the same or similar ways over and over again. Despite the fact that great pains have been taken to define repetitiveness in movements, there is really no definitive answer as to what is the best method. This is primarily because there are two different expressions relating to repetitiveness: (1) frequency of movements and muscle activity, and (2) similarity between repeated events. These two are sometimes confused. The first relates to the degree of change in loading per unit of time. Similarity describes how similarly a repeated movement is performed over time. The performance can be measured in parameters such as time taken, force required, and the range of the movement in space.

5.4 PHYSIOLOGY OF REPETITIVE MOVEMENTS

5.4.1 HOW DO WE CHOOSE TO MOVE?

Movements of various parts of the body can arise in many different ways. We can choose to perform movements ourselves, movements may occur through reflexes, or be governed by more automated systems such as when we walk, for example. We can also be subject to external forces that set our body parts in motion. When a movement occurs, it is often a combination of different ways of moving a body part. Reflexes and automated movements can be affected by external forces, and by our own will to move a part of our body; for example, we can counter changes in the ground we are walking on or change our walking style. We can also compensate by increasing the force we generate, for example, when it is difficult to cut through something hard.

The repetitive movements that Janis carries out when he is working are movements that he himself initiates and over which he has conscious control. To carry out these movements his nervous system needs to know and analyse a number of different factors affecting his performance, for example, what Janis' body looks like and how it works, what the surroundings are like, what emotional condition Janis is in, his motivation in performing the movements and previous experiences of doing so. In Janis' case, the environment provides important limits to how movements can be performed. Those cuts he makes must be correct and must be made within a specific predetermined time, so that Janis manages to do as much as possible during his working day. This therefore governs what opportunities Janis himself has to choose regarding how he moves. Janis is also limited by the fact that he is holding a sharp object (the knife) and in certain cases has to make very specific movements to avoid injuring himself.

5.4.2 INTERNAL MODELS AND MOVEMENTS

In order to be able to perform the movements that are required and at the same time stay within the limitations imposed by the environment, information is necessary about what the environment looks like and its properties, and what the body looks like and how it functions. The dominant theory for how the central nervous system (CNS) can deal with all this information is based on the creation of internal models with regard to how the movement should be performed [Wolpert et al. 1995]. The internal models describe the properties of the environment and the body, for example, how our arms work or the properties of the knife and its function. On the basis of these models, the nervous system can predict what is going to happen if, for example, the muscles of the forearm are activated when the hand is holding a knife. In order to create accurate models of the body, the objects we hold, and the environment, information from the eye and different receptors in muscles, joints, and the skin are very important. The creation of internal models is the one of the first steps in performing a movement.

In order to build up internal models, different reference systems are created. These may either be centred on ourselves or on an object outside our body. A reference system also helps determine how various things relate to one another. For example, one reference system describes where the knife in my hand is located with reference to the eye that is, looking at the knife. The body then creates a number of different reference systems that are linked together. As we know how the eyes are located in relation to the head, a reference system between the head and the eyes can be created. This is developed further, as we know how our arms are placed in relation to our head and our arms to our hands and so forth. Finally, an entire reference system including the head, the arm, the hand, and the knife is built up. The knife can now be related to our own body and the environment around us. As these various reference systems are based on each other, they are also dependent on each other, and if one reference system changes, the others must be updated. Previous experiences of handling a knife help us to create and update the reference systems dealing with objects like the knife more quickly and thereby also updating the internal models. The CNS works constantly with a set of different reference systems and makes changes to these. With the aid of the internal models that have been built up of ourselves and our environment, a specific internal model is created for a limited movement we intend to perform. The internal model forms a basis for determining how and in what order muscles should be used, a "motor pattern," which is explained in the next section.

5.4.3   Motor Patterns

Signals that activate muscles are called motor commands and are generated in different areas of the brain. Several motor commands build up a motor pattern, which in turn is dependent on what the internal model of the movement looks like. A motor pattern can be seen as the solution the internal model uses in order to achieve its aim with a movement. The choice of motor pattern is limited in this way not merely by properties in the environment but also by the object that has to be handled and by the properties of the musculoskeletal system. Lifting something heavy with a particular movement does not produce the same motor patterns as lifting something light in the same trajectory. Previous experience of making similar movements is of importance for which motor pattern will be used and therefore for the performance of the movement. When an internal model is established, it becomes easier to find motor patterns when similar movements are to be repeated.

When Janis makes his repeated cutting movements, he does not need to create a new motor pattern each time, but can use the same pattern time and time again, possibly with some minor adaptation, as there is an internal model for the movement. However, neither the internal model nor the motor pattern used are constant, but may change with the help of information generated by receptors in muscles, joints, and the skin. This results in a tendency for repetitive movements to imitate the last movement carried out, and not the first in a series of repeats. Janis' cutting movements do not, therefore, need to be as similar in each repetition as they may appear at first glance.

The nervous system can also generalize between two similar movements where internal models support each other. In this way, no new motor patterns need to be created all the time just because a minor modification occurs, for example, in the environment. For any internal model to become permanent, a certain amount of time is needed for confirming the motor memory [Cohen and Robertsson 2007]. It is important during the period of confirmation not to create internal models that are similar, as this makes confirmation more difficult. Gradual changes in the task lead to the internal model being gradually changed, and somewhere there is a limit where the modified model differs from the original to such an extent that it is stored as a new model.

5.4.4   Degrees of Freedom and Redundancy in Motor Systems

Every joint in the body can move in a number of directions. The shoulder joint can, for example, be angled upward, downward to either side, and can rotate. These movements are independent of each other and can occur without affecting each other. The number of possible independent movements that can be made in a joint is called the number of degrees of freedom of the joint. The arm consists of hand, elbow, and shoulder joints, and these joints together have a number of degrees of freedom. Put simply, the wrist may be said to have two degrees of freedom—movements upward, and downward, and to the right and the left. The bones of the forearm, which are linked to the wrist and elbow, can rotate and in this way produce rotation of the wrist and hand. This rotation adds one degree of freedom. The elbow joint adds a further degree of freedom, and the shoulder joint adds three degrees of freedom. In total, the whole arm in this simplified model has seven degrees of freedom. In order to determine the position of the hand, only three degrees of freedom are required, corresponding to the three dimensions in space. There is, therefore, a surplus of degrees of freedom for positioning the hand in space using the arm, and this surplus is called redundancy. Redundancy means that there are several different possibilities for combining angles of the shoulder, elbow, and wrist joints and still produce the same position for the hand.

In order to achieve these different angles of the joint, different motor patterns are needed. When Janis cuts his shoulders of pork, the redundancy with regard to degrees of freedom means that he will be able to do this in many different ways and still make all the joints of meat the same. What causes Janis to use one particular motor pattern rather than another is not clear. Nor is it certain that Janis makes use of different motor patterns during the working day. He may perhaps be using the same one the whole time. It is not just the performance of the movement that can be varied; the activation of individual muscles can also be varied. In trials where people have had to hold their arms stretched straight out from their body and are therefore loading the neck muscles, different people have differing abilities to change their motor patterns and at the same time reach the desired goal as regards muscle activation. It is not clear as to what significance this has for any potential development of disorders. In an experimental study, experienced, healthy butchers and a control group of healthy "nonbutchers" carried out a simulated cutting task consisting of a series of cutting movements [Madeleine et al. 2003]. The two groups demonstrated different activation patterns of the muscles involved, where the experienced group was characterized by a more varied pattern and lower activation levels. This illustrates that they may have different ways of activating groups of muscles, but nevertheless achieves the same result. Even within a muscle there are degrees of freedom. A large number of different combinations of motor units can lead to the same final effect of muscle activation. When Janis carries out a repetitive task, his body therefore provides the preconditions for varying both motor pattern and muscle activation in different ways, despite the fact that the aim of the movement is the same.

5.4.5 RESTRICTIONS ON THE UTILIZATION OF THE NUMBER OF DEGREES OF FREEDOM

How the number of degrees of freedom is used may be restricted by factors from the environment. For example, movements with considerable demands for precision can result in a smaller number of degrees of freedom being available, as the entire movement has to be carried out in a very exact way in each work cycle. Some of the combinations of joint angles which theoretically lead to the same hand position are not "allowed" in practice. In this way, there is less possibility of variation of muscle activity between different cycles. This is particularly important in jobs where there is repetition of a movement and at the same time a requirement for precision. We also have a built-in restriction on how different degrees of freedom can be used: how the joint surfaces and ligaments are designed in different joints. These structures only allow certain movements and thereby restrict how the degrees of freedom can be used. The utilization of the degrees of freedom is therefore also affected if any structure, for example, a joint, is damaged. Adaptations in motor function may then lead to certain muscle activities being impossible to perform without resulting in pain. Pain in itself is also a factor that influences the internal model and thereby the choice of motor pattern.

5.4.6 MUSCLE SYNERGIES

An important property of the body which contributes to motor patterns being variable is that there are more muscles around most joint than are theoretically required to carry out all of those movements permitted by the anatomy of the joint. The CNS in this way has to choose suitable patterns of muscle activations that reduce the redundancy in the degrees of freedom to carry out the specific movement desired. A suitable pattern of this kind is called a muscle synergy. Different muscles therefore work together in synergies in order to achieve a certain desired mechanical effect. For example, to simplify somewhat, the head has three degrees of freedom in relation to the body, but there are 23 muscles on each side of the spine that participate in moving the head. Depending on the desired movement, the CNS chooses the synergy that best suits the task. For example, moving your gaze from one place to another is carried out in three different ways:

- Eye movements
- Head movements
- A combination of eye and head movements

How synergies are organized precisely and governed is relatively unknown. Presumably, some aspects are innate and others are acquired through learning. As learning is important, different people may have learnt different methods to coordinate their muscles in synergies when they perform a particular movement. The different ways of working are also revealed through different cycle times, force levels, and a different number of cuts, for example, on the part of butchers carrying out the same task. The butchers whose muscle activity was measured may have changed their motor patterns to better adapt to the job. In a comparison between experienced meat dressers and beginners, the experienced butchers' muscle activation patterns resulted overall in a shorter period for the work cycle, something that may be of financial advantage, as butchers often work on individual piece rates.

Coordination of the muscles in the synergy is well defined to specific movements. Many muscles that act across several joints may be agonists (working in the same direction) in one case and antagonists (working in the opposite direction) in another, depending entirely on what types of movements are being carried out. In movements that comprise several joints and muscles acting across both one and two joints, a movement becomes a pattern of muscle activations in which muscles can work together with each other in certain components of the movement and against each other in other components. In complicated movements, the coordination of activation and deactivation of agonists and antagonists is of major significance. If synergies are not adapted to a movement, there is a risk that the aim of the movement is not achieved, and that strain during the movement increases as muscles counteract each other. There can, however, be conflicts where one and the same muscle is needed for different tasks. One muscle that is activated in many different tasks is the trapezius muscle. This muscle generally helps in stabilization, and in movements of the shoulder joint or the head. The trapezius muscle is divided up into different functional units, which makes it easy for the muscle to take part in different tasks at the same time. When Janis cuts out shoulders of pork, he is dependent on muscle synergies functioning optimally. A disturbance of, for example, the trapezius muscle might lead to a failure of the work apportionment within the muscle and certain portions of the muscle becoming overloaded.

An important tool enabling the body to evaluate and monitor a synergy is sensory information. Sensory information affects all the stages in the chain from planning to choice of motor pattern and performance of a movement. Moreover, the sensory information is important for optimization and adaptation of muscle activity.

5.4.7 Sensory Motor Function in Repetitive Movements

There is a constant interaction between sensory and motor signals in the nervous system. The performance of movements generates sensory signals that affect how future movements will be carried out. Sensory signals into our CNS come from receptors placed in our muscles, joints, and skin. Visual impressions and those signals sent by the organs of balance in the ears are added to sensory signals. Common to all signals is that they are continually monitored and analysed by the CNS. All sensory signals do not reach our consciousness, but sometimes remain at a deeper and more unconscious level. Sensory information from the periphery, that is, from muscles, joints, and skin, is sent through the spinal cord via different switching stations, and ends up in the somatosensory areas of the cerebral cortex. On the way up to and in the somatosensory areas, sensory information is collated from various sources into an overall picture. Much points to the fact that the CNS can weight sensory information—that is, to ascribe varying significance to sensory information from different parts or sources. There are strong links between somatosensory and motor areas in the cerebral cortex. There are also other places where sensory and motor functions cooperate, for example, in the cerebellum and at different levels in the spinal cord. Sensory information is in this way particularly important in the learning of new movements, that is, the creation of new internal models and motor patterns. The principal sources of sensory information for movements are vision and the muscle spindles. Muscle spindles are 1–2 mm contractile receptors that lie in parallel to ordinary muscle cells in most of the muscles of the body. The number varies between different muscles. Some muscles have no muscle spindles, while others have thousands. Muscle spindles send continuous information about

how tense various muscles are and whether they are being lengthened or
shortened.

Muscle spindles are therefore a great help in the CNS's understanding of movement and positions in the musculoskeletal system (see also Chapter 6). The importance of sensory information from muscle spindles depends on the context in which
the movement is performed and how the information is weighted. When Janis makes
an automated rhythmical movement, for example, such as walking, information from
the muscle spindles plays a minor role. There are already clear internal models used
by the nervous system to predict what is going to happen in precisely these movements. This sensory information is used more for controlling starting points and
ending points, and for being part of how the interplay between different muscles
works in motor patterns at the spinal cord level. Sensory information can then be
used as signals to change to between different motor patterns. In carrying out an
automated rhythmical movement, it is only in cases of unforeseen deviations from
the anticipated movement when the sensory information is used as a basis for direct
correction of muscle activity. How substantial this correction is depends on the
weighting that the nervous system places on the signals. The significance of certain
signals may, however, be altered by the nervous system, if the signals are assessed as
containing important information.

Sensory information from muscle spindles and other receptors responding to
movement, for example, in joints and skin, affects motor functions through reflexes
at the spinal cord level. The most classic reflex which a muscle spindle is involved in
is the stretching reflex, where stretching a muscle spindle through direct connections
to the α-motor neuron leads to a contraction of the same muscle. This means that
activity in the muscle spindle can make it easier for the α-motor neuron to send signals out to the muscle. Just as the muscle spindles in themselves are not identical, nor
is their influence on the α-motor neuron completely uniform. The muscle spindles
have been shown to help in controlling which motor units are to be activated and
deactivated [Grande and Cafarelli 2003]. In muscle fatigue, the sensory information
from the muscle spindles changes [Pedersen et al. 1998]. This has consequences for
all motor functions dependent on sensory information. Above all, the internal models
that use sensory information to predict how the body and the environment will react
are affected. If the sensory information is disturbed, it will affect the internal models
and, by extension, which muscle synergies are used to perform a movement.

5.4.8 DIRECT RESPONSES IN EXPOSURE TO HIGHLY REPETITIVE MOVEMENTS

Exposure to highly repetitive movements leads to a number of direct responses in the
body. These responses consist of reactions in muscles, joints, tendons, and the nervous system to the load to which the body is subjected, including the muscle work
that is being carried out. As highly repetitive work implies recurrent activation of
muscles in arms, shoulders and neck, with momentary activations above 50% of the
MVC, muscle fatigue is a natural and anticipated response. Acute muscle fatigue
means that the muscle loses some of its ability to develop force. This, in turn, means
that a greater mobilization is required on the part of the CNS for that muscle to cope
with the same task. It is very rare in working life that development of muscle fatigue

in individual muscles continues until the work can no longer be carried out. Instead, the aim of the movement is realized using different muscle synergies. Either a modification occurs in the motor pattern, or a new internal model and a completely new motor pattern are created. Muscle fatigue is often accompanied by aches and a warm sensation in the muscles which comes from muscle work, as well as a general feeling of fatigue. At the end of the work a feeling of stiffness can also appear, and pain when moving the muscles are not uncommon. As acute muscle fatigue also affects force development in a negative way, muscle weakness can be experienced several days after the end of the work. In muscles whose primary task is to stabilize the body so that the arms are able to carry out comprehensive repetitive movements, acute muscle fatigue is not as pronounced, presumably because the activation level is lower. In different types of repetitive force load, force levels and cycle times for variations of force have different effects on muscles in the forearm and muscles in the neck.

The pattern for muscle fatigue in the trapezius muscle during continuous work differs a great deal between different individuals, while the pattern for muscle fatigue in the surrounding muscles are more similar for different individuals. This difference in the development of fatigue in the trapezius muscle in particular presumably has to do with the role of the muscle both as a stabilizer for the head and in moving the shoulder joints. The development of fatigue is also sensitive to variations in load. Experiments show that the experience of fatigue occurs significantly later if the load is varied than if the load is constant, even if the total load intensity does not change [Mathiassen 1993]. The experience of fatigue is counteracted by variation in load, and in this way the work may be carried out for a long period. When Janis is working, he is exposed to a mixture of constant and dynamic load, which means that the development of fatigue becomes complicated.

In repetitive movements, biochemical changes in the muscle cells occur. These changes are the consequences of the way in which the muscle works. An important precondition for the occurrence of a muscle contraction is liberation of calcium ions in the individual muscle cell. In the prolonged repetitive activation of a motor unit, there is a risk of the structures in the muscle cell responsible for calcium release becoming damaged. If damage occurs, it becomes more difficult for muscle cells to contract. This makes for very prolonged muscle fatigue with a reduction in force development lasting up to 72 h after the end of the work. As the effect of muscle fatigue is prolonged, there is an impending risk that the period for rest and recuperation will be insufficient. Recuperation is important for the muscle to be able to repair itself. If the repair cannot be carried out, there is a risk that motor units that have to work hard will become damaged.

In repetitive muscle work and with loads without any change in force and muscle length (so-called static load, see Chapter 6), lactic acid (lactate) is formed as a consequence of a reduction in oxygen supply to the muscle. Only 20–60 min of repetitive arm work at 10–15% of MVC produces an increase in lactic acid in the trapezius muscle. There is more about the production of lactic acid and its effects in Chapter 2. Another substance that is produced in muscle contractions when muscles are tense without any change in force and position is bradykinine (BK). Production also occurs as a consequence of changes in muscle pH which in turn may arise if a great deal of lactic acid is formed. BK affects nerve receptors in the muscle so that they become activated and acquire increased sensitivity. The greater sensitivity applies not merely to lactic acid and BK, but also to other substances which may be produced, and to mechanical and thermal stimulation. Other substances that may be formed are arachidonic acid and various interleukins. Arachidonic acid forms the basis for the production of prostaglandins. Preventing the production of prostaglandins is the aim of certain painkillers belonging to the group known as *NSAIDs* (see Fact Box 5.1). Interleukin 6 (IL-6) is important for repairs and building up muscle, at the same time as it is also involved in inflammation and pain.

FACT BOX 5.1

NSAIDS (NON-STEROID ANTIINFLAMMATORY DRUGS)

Analgesic drugs containing, for example, acetylsalicylic acid, ibuprofen, or naproxen. Drugs containing paracetamol do not belong to this group.

Common to all direct responses in work is that they contribute to rebuilding muscles and other tissues. Direct responses therefore need not only be regarded as unpleasant or troublesome, but are a precondition for the slow adaptation of the body to various loads, for example, greater muscle capacity as a result of high muscle activity (see also Chapter 6). For the body able to rebuild itself rest is needed and thus recuperation after work becomes important.

Repetitive muscle activation also leads to tendons and joints being affected, and in the same way as muscles, these need a period of recuperation after work in order to be repaired and reinforced. Tendons and joints have fewer blood vessels to help to carry away the waste products formed and to provide them with new energy and new building blocks for repair. If only a small part of a muscle has to work hard, then a small part of the tendon also receives greater load. Generally speaking, it takes an even longer time for a tendon to build up its strength than it does for a muscle. This means that the tendon for certain periods may be disproportionately weaker than the muscle. Often in the transition between tendon and muscle there is a sensitive point, as the tissue is weaker there. Nerves can be damaged, for example, by repetitive movements of the wrist. What is most common is that an inflammation forms in and around the nerves, and as a consequence the nerves become compressed. This leads, in the longer term, to pain, deterioration in motor function, and loss of sensation in the area served by the nerves. An example of a condition of this kind is carpal tunnel syndrome.

Studies of direct responses to repetitive work also show that our movement and position sense are affected, presumably because the ability of the muscle spindles to provide good information deteriorates (see Fact Box 5.2). Deterioration in movement and position sense produces less well-controlled movements, as planning and performance of movements are dependent on good sensory information. This may be linked to a feeling of clumsiness and heavy-handedness.

FACT BOX 5.2

MOVEMENT AND POSITION SENSE

This means that we can keep track of where our arms and legs are without needing to use our eyesight. Movement and position sense is very dependent on sensory information from the muscle spindles, but information from receptors in tendons, joints, and skin is also significant.

5.5 PATHOPHYSIOLOGICAL MECHANISMS IN EXPOSURE TO HIGHLY REPETITIVE MOVEMENTS

Highly repetitive work is a combination of dynamic work requiring force, using the arms and part of the shoulder, and a prolonged activation of the muscles in the neck and shoulder at a lower level of force. For a detailed review of the explanatory mechanisms for pain from prolonged low-intensity load, see Chapter 6.

Fatigue has long been considered to be a preliminary stage to injury. The current recommendation that breaks at work are good for preventing disorders is based partly on the fact that subjective fatigue is lower if a job is divided up into smaller parts with breaks between than if the same job is carried out without a break. It is, however, important to take into account the fact that physiological changes in the muscle do not always go hand-in-hand with the subjective fatigue experience. For example, the levels of potassium outside the muscle cells do not always decrease in parallel with subjective fatigue during a break. Potassium remains high during the break even though the subjective fatigue diminishes. If the work continues after the break, the muscles will not have rested sufficiently, even though it feels as if it has. The potassium level is linked to the level of calcium, which is very important for the muscle to function in an efficient way. As the breaks do not provide a sufficient reduction in the potassium level, there is therefore a risk that the muscle might become damaged as a result of changes in calcium levels.

The exact mechanisms for how disorders arise in an exposure to repetitive movements with a requirement for force and precision are not known. There is, however, a great deal of data to support the fact that control and allocation of muscle activity are very important.

5.5.1 CHANGES IN WORK ALLOCATION IN THE MUSCLE

When a muscle is activated, the motor units start up in a particular order. Exactly which motor units are used is governed not merely by what force is required and how rapidly the force needs to be developed, but also how the movement is to be performed in space. Normally, an assemblage of motor units and muscle synergies is used in a movement. In order to be able to use different muscle synergies and vary the motor units, a surplus of degrees of freedom is required. If the possibility of making use of the number of degrees of freedom is restricted, the opportunities for variation are reduced, and the risk that the same motor units or muscles have to perform a large share of the task increases.

When Janis is working, he is exposed to constant strain on his neck muscles at the same time as the more forceful repetitive work with his arms. The strain on his neck muscles is produced partly by the need to stabilize his head, and also by reaction forces to the work in his forearm. With prolonged low-intensity loads, there is a risk that the motor units activated first are also those that are deactivated last. There is then a risk that the muscle cells in these motor units will not recover sufficiently. These motor units are called Cinderella units, as they are the first to start work and the last to finish. When registering electrical activity from individual motor units in individuals both at work and in an experimental environment, continuously active motor units have been registered for ~1 h. There is, however, still some uncertainty as to what significance these Cinderella units have in the development of disorders [Hägg 2000]. More about the possible significance of the Cinderella units in prolonged low-intensity loads can be found in Chapter 6. The risk of Cinderella units forming in Janis' arm muscles may be assumed to be lower than in his neck muscles, as dynamic movements with more powerful muscle contractions seem to make it easy for any Cinderella units to be replaced by other motor units. Janis' work should then give him an advantage in his arm muscles, as the allocation of work in those muscles should be better. There are, however, other factors in the work which act in the opposite direction. The precision needed in the work restricts the number of degrees of freedom in Janis' movements. For Janis, there is only a small opportunity of varying muscle activations if he is going to cut out perfect joints of meat, which his performance-related pay is partly based on. This can lead to the same motor units having to work in Janis too, as there are a limited number which can perform a specific task that Janis carries out. The need to see clearly also reduces the opportunity of exploiting the degrees of freedom, as a consequence of holding his head still is an activation of neck muscles that reduce the opportunities of changing the position of his neck and shoulders. This happens independently of whether Janis has otherwise perfect vision. The neck muscles find themselves in a complicated situation, where the requirement of stability of the head has to be combined with the need for movement in the shoulder. Sensory information here plays an important role in creating suitable activation patterns to meet both demands.

Janis employs considerable force in his movements, which may lead to extensive muscle fatigue in the motor units which take the greatest share of the load. When muscle fibres in a motor unit begin to tire, the activation frequency of the motor unit increases. The greater activation frequency and the local changes in the muscle cells contribute to the feeling of fatigue in the muscle. However, it is not certain that the feeling of fatigue is relative to the load to which individual muscle cells are subjected, rather than fatigue reflecting the condition of the entire muscle. When certain motor units cannot generate the desired force, other motor units have to be activated to be able to carry out the task. These other motor units do not need to be optimal for this specific task, but there is a risk that this leads to greater loads. The greater loads derive from a greater element of simultaneous activations of counteracting muscles (antagonists) in order to stabilize and maintain the precision in the development of force. An increase in the agonist–antagonist contraction in muscle fatigue can lead to a decrease in the time of individual muscles for rest and to an increase in the production of various chemical substances in the muscles.

Simultaneous activation of antagonistic muscles (agonists and antagonists) around joints is also a way for the CNS to increase control over a movement. Weaker control of movements can result from repetitive movements affecting the movement and position sense. The sensory information that builds up the movement and position sense is important in creating suitable internal models. If the models are not adapted to what is actually happening in the tissue, there is a risk that the wrong motor patterns are activated. For Janis these incorrect motor patterns can mean that the performance of repetitive movements in combination with force and precision requirements leads to a greater load on his muscles, tendons, and joints.

The reduced opportunity of relieving tired motor units and/or muscles generates a potential starting point for disorders. It has, however, been shown that different people differ in how well they make use of the opportunities that exist to carry out the same work with different movement and muscle activation patterns. To what extent this predicts who is going to develop disorders or not has currently not been demonstrated, but it is a reasonable hypothesis. Muscle pain in itself affects motor function by changing muscle synergies with the main aim of reducing the load on the painful muscle. If the aim of the movement is maintained, despite the fact that the main muscle for this movement is signalling pain, it will be necessary for other muscles to compensate for any loss of force. Preexisting pain can therefore further increase the load on other muscles. This can also be seen as a pure defense mechanism to protect a potentially damaged muscle from further damage.

A prolonged activation of individual motor units or the entire muscle generates production of various metabolites and inflammatory substances. The production of metabolites is primarily an effect of the need for energy in the muscle cells. Inflammatory substances are produced as a response to the muscle cells being damaged and needing to be repaired. Repair, however, requires rest. A constant state of insufficient recuperation and overconcentration of metabolites and various inflammatory substances leads to an increased and altered sensitivity on the part of sensory nerves in and around the muscle. As these nerves are not only sensitive to various substances, but also to mechanical influence, the threshold will be lowered for when mechanical influence results in pain. If a prolonged activation of sensory nerves continues, physical changes in the connections in the spinal cord may occur. These changes in the connections may lead to changes in the CNS interpretation of signals from the periphery. For example, sensory nerves, which do not normally transmit signals about nociceptive pain, now are interpreted by the CNS to do so. Generally speaking, an increase has occurred in how sensitive the nervous system is to activation. This increased and altered sensitivity is called central sensitization. Central sensitization may be an important mechanism, alongside physiological, psychological, and social risk factors, that develop transitory episodes of pain and discomfort into prolonged chronic pain. Central sensitization may also be an important source for why pain in the muscle continues when the original causes of the pain are removed. As regards the origin of chronic pain, see also Chapter 6.

Experimental studies on prolonged exposure to repetitive loads in human beings are rare. However, there are data from animal studies. In a series of experiments on repetitive work, rats were trained to receive food at a particular rate [Barbe and Barr 2006]. The rats performed a special movement four times a minute for 2 h/day and

3 days/week for a total of 8 weeks. After only 3 weeks the rats showed marked increases in inflammatory substances and changes to their motor function. The production of inflammatory substances increased over the first 6 weeks of load, to subsequently decrease somewhat up to the eighth and final week, but not back to their original value. The experiment shows also that cells from the immune defense system had been activated and had migrated into muscle and tendon tissue.

5.5.2   TENDONS AND NERVES

Repetitive movements not only affect muscles, but also tendons and nerves. How tendons are loaded depends on which muscle activation is occurring. There is no opportunity for variation of load within the tendon; rather the tendon is exposed to the variation produced by the muscle. The actual variations in load within the tendon are, however, less than in the muscles, and the risk of overload of the tendon is greater. In this way, the tendon is in greater need of rest to recuperate after a load. It is therefore probable that disorders often appear first in tendons and subsequently in muscles. Damage to tendons may probably arise because of a number of different mechanisms, for example, mechanical wear with resulting inflammation and changes to the structure of the tendon, or a reduction in the blood flow to the tendon. Both cases lead to structural changes in the tendon which mean that the susceptibility of the tendon to strain is increased. Reduction of blood flow is due to compression of the blood vessels serving the tendon. Tendons have from the outset relatively poor blood flow, which contributes to the increased sensitivity.

Nerves, too, can be affected by repetitive movements insofar as nerves slide against the surrounding tissue when they change position. A constant repetition of the same movement leads to an inflammation process starting in the epineurium with consequent swelling. Swelling causes increased pressure on the nerve with pricking sensations, numbness and pain as a result. In the healing process there is a risk that the nerve's ability to slide in the epineurium is reduced, which makes for increased sensitivity to strain in the area.

5.6   CONSEQUENCES OF REPETITIVE LOADS AMONG THE POPULATION

In Europe, 49% of workers in the groups of machine operators and assemblers stated that work affects their health, and 34% report muscle disorders [Eurofond 2007]. A study from Sweden shows that during the period 1996–2005 the proportion of all employed women stating that work has caused disorders as a result of short repeated operations has been around 4%, with a peak in 2003 at 4.9% [SWEA 2008]. For men during the same period an increase from 2.6% in 1996 to 3.4% in 2005 can be seen in disorders relating to short repeated operations [SWEA 2008]. Between 2000 and 2005 almost 5% of men and a good 9% of women within manufacturing industry stated that they had disorders as a result of short repeated operations; no details of the disorders were, however, given [SWEA 2008]. Among butchers, meat-dressers and vehicle assemblers the frequency of reported work-related diseases because of physical strain was more than 15 cases/1000 individuals in 2004 [SWEA 2005].

In Denmark during the period 1993–2003 the average number of reported cases per year of physical symptoms related to repeated monotonous work was a good 11 cases/1000 employees, among pig and cattle slaughterers [DWEA 2003]. The specific diagnosis of supraspinatus tendinitis, that is to say, inflammation in one of the tendons of the shoulder, occurred in almost 9% of slaughterhouse workers in Denmark in the same period [Frost et al. 2002]. Previously, it has been reported that up to 14.5% of workers in fish processing have epicondylitis [Chiang et al. 1993], and up to 12% of workers in other jobs with high demands for repetitiveness [Kurppa et al. 1991]. The effect on the median nerve, known as carpal tunnel syndrome, is a well-known result of repetitive work (cycle times of <10 s), particularly in combination with forces of 4 kg or more [van Rijn et al. 2009]. There is considerable support for the fact that repetitive movements under strain are a strong risk factor for the development of muscle pain and inflammation in tendons and tendon attachments in the thumb, wrist, elbow, and shoulder.

For muscles in the neck/shoulder, repetitive work, along with a prolonged, stooped head position, has been identified as risk factors for developing pain. Repetitive work in combination with prolonged so-called static load (loaded with no changes in force or muscle length) in the neck generally produces a greater risk of disorders.

5.7   RISK ASSESSMENT

In order to identify risks in connection with repetitive work, it is important to study carefully the work being done. Initially typical tasks are identified at work, particularly those tasks where repetitive operations occur over a long work task period. In Janis' case, he lifts down a large part of a pig and subsequently works on it. He does this several times during the workday. Several different work cycles can be identified in this task. Janis begins by cutting out a bone and subsequently cutting the meat into a specific joint. On this basis, a description on the classification of risk factors for each work cycle will be made, for example, how repetitive the work is, what force is required, what the work posture is like, demands on precision, and other factors such as cold. The different work cycles are brought together and the order in which they are carried out, how long each work cycle lasts, and time allocated for recuperation or rest is taken into account. Generally speaking, in a risk assessment of repetitive work it is important to study the combination of repeated movements, force development, duration, and the need for precision. It is important to take into account how long the exposure occurs, for the individual risk factors on the one hand and for combinations of risk factors on the other. There are three main groups of methods to assess exposure and by extension risks:

- Self-reporting
- Observations
- Direct measurements

Self-reporting includes methods such as interviews, questionnaires, and diaries. Employers may assess the incidence of, for example, a particular work posture, frequency of movements, force development, duration of certain work tasks, or fatigue.

The advantage is that the methods are often simple to use and relatively inexpensive. Usually, a large number of participants are needed for the data collected to be representative of the group being investigated. In using observational methods, various parameters are annotated by an observer on preprinted forms or entered directly into a computer. What is annotated or entered varies between different systematic methods; for example, some only take into account individual parts of the body and others take into account whole work postures. The advantage of simple observation methods is that they are relatively practical and inexpensive, and highly suited to investigating work postures and simple repetitive tasks. The disadvantages are that the methods can be influenced by the fact that the same observer assesses different people at different time points during the working day, and different observers do not assess in the same way. Video recording and subsequent analysis of the film are more time consuming, and are used more for jobs with a more varied pattern of movements. Video-based systems involve high costs and require technical support and user training.

Below, a selection of observation methods is summarized, as well as their main areas of use and function. For additional methods and a short description of these, the reader is referred to the publication by David GC (see Further Reading).

Rapid upper-limb assessment provides a quick impression of primarily work postures and external load. The method not only makes use of classifications of work postures, but also provides information about repetition and the use of force and is usable as initial screening of all jobs where there is exposure to repetitive operations.

The strain index is a semiquantitative method in which six different variables relating to the task are measured or assessed. All the variables are divided into five criteria, where each criterion corresponds to a figure. All the assessments are multiplied, and this produces an index. Studies have shown that this index measures the risk of disorder.

The occupational repetitive actions (OCRA) Risk Index is used to assess work postures, force, and cyclicity in repetitive work. The method takes into account many different variables, which are weighted together in a risk value.

Direct measurement methods are based on the fact that different measuring instruments are attached to the individual who performs a task. This may be equipment to measure joint angles, range of motion, muscle activity, or forces. This method is suitable for studies of simulated work tasks in controlled environments such as laboratories. It is, however, quite possible to take measurements at workplaces, but this requires experience and great knowledge. When attaching measuring equipment to individuals, one also risks affecting their method of work.

5.8 WHAT MEASURES CAN BE TAKEN TO MINIMIZE THE RISKS OF REPETITIVE WORK?

An increased variation is prescribed as one of the most important measures to counter risk in critical conditions involving repetitive work. Variation of work tasks and variation within the work task are two types of variation that are conceivable. Also the use of breaks for recuperation is part of a good intervention: on the one hand, the breaks give greater variation; on the other, they provide a direct opportunity for recuperation. It is also important for there to be sufficient time for recuperation between shifts, so that the body is as rested as possible at the start of work. Interventions in the work that lead to additional degrees of freedom being available for the motor system are presumably also valuable. This can be done through workers being allowed to rotate between many different work tasks, what is known as job rotation (see Chapter 6, Section 6.14). It is important that the various tasks in a job rotation are not too similar to each other, because then there would be no variation in the exposure. A good tool is to build up an organization where disorders are caught early through systematic risk assessment, and where there is follow-up of the work situations as well as close collaboration between occupational health care and the companies.

5.9 WHAT DO LAWS, REGULATIONS, AND PROVISIONS HAVE TO SAY?

Specific legislation on highly repetitive work does not exist either at the EU level or at the federal level in the United States. Within the EU there are a number of different directives regarding exposure to risk factors in working life, but none that directly address highly repetitive movements. On the other hand, in the EU there is an overarching framework directive which applies to measures for promoting improvements in employees' safety and health at work (Directive 89/391/EEC). In this directive there is an explicit reference in Article 6 which in part is applicable to highly repetitive work. This is based on the fact that the employer is bound to adapt the work to the individual, in particular as regards the design of workplaces, choice of work equipment, and choice of work and production methods, with the intention of primarily reducing monotonous work and work with fixed piecework rates, and reducing their effects on health. In the United States there is also comprehensive legislation about worker safety and health which includes the fact that the employer is responsible for removing risks at work which may cause serious injury to the worker.

In many European countries there are, however, more detailed national rules and advice as well as regulations and support concerning highly repetitive movements. In Sweden, the texts of the provisions are detailed as regards highly repetitive work [SWEA 1998]. For example, the paragraph dealing with monotonous, repeated, strictly controlled, or fixed work runs: "the employer must ensure that work that is repeated monotonously, strictly controlled or fixed does not normally occur." It further states that if special circumstances demand that such work nevertheless must be performed, the employer should take measures to minimize the risk; for example, "job rotation, job enlargement, breaks, or other measures to increase variation at work." In the general advice clarifying the provisions in Sweden, the point that carrying out repeated movements often requires static work in the surrounding muscles is stressed. There is, therefore, a link between different forms of physical load. The general advice also describes favourable load. It is characterized by repeated variation, balance between activities and recuperation as well as time limits. It is also pointed out here that tiring physical strain is not necessarily injurious to health, but that exposure times and recuperation period are important factors to assess. There are also standards applying to a highly repetitive work published by the International organization for standardization (ISO 11228-3 2007) and European Standard (CEN prEN1005-5 2007) which provides guidance in the identification and evaluation of risk factors associated with highly repetitive work. Standards are, however, only recommendations and not legislation.

5.10   SUMMARY

Highly repetitive jobs often comprise a combination of repetitive movements, force requirements, and precision. A difficult job is that of a butcher, but employees in other professions, for example, in the food production and the assembly industries, also carry out highly repetitive work. In order to perform a movement, our brain needs to collect information before the movement is performed, plan the movement, carry it out, and finally evaluate the movement. In order to do this, the brain creates an internal model of various conditions for the movement. Then a suitable motor pattern is generated, moving, for example, our arm to the desired location. There are often a number of ways of performing a movement and still achieving the end result. Restrictions on the ability to vary a movement can result from demands for precision or force, or that the repetitive movement has to be carried out in a particular way. The restriction may lead to overuse of particular muscles and/or muscle fibres if muscle synergies and motor units cannot be alternated. The lack of variation becomes most prominent when the repetitive movements, demands for forced, and demands for precision are combined. Exactly how this subsequently leads to pain has not been clarified, but presumably a long-term accumulation of various substances produced in the muscle resulting in a central sensitization are important. Pain in itself also result in the body changing its method of moving. This may, in turn, lead to overload on other muscles or muscle fibres. An activation of muscles in a very similar way also provides a potential effect on tendons, as the repeated load on the muscle is transferred to the tendon. In order to counteract disorders, an increase in variation both between work tasks and within work tasks as well as sufficient time for recuperation is suggested.

REFERENCES

Barbe, MF., Barr, AE. 2006. Inflammation and the pathophysiology of work-related musculo-skeletal disorders. *Brain, Behavior, and Immunity* 20:423–29.

CEN prEN1005-5. 2007. *European Committee for Standardization*.

Chiang, HC., Ko, YC., Chen, SS., Yu, HS., Wu, TN., Chang, PY. 1993. Prevalence of shoulder and upper-limb disorders among workers in the fish-processing industry. *Scand J Work Environ Health* 19:126–31.

Cohen, D., Robertson, E. 2007. Motor sequence consolidation: Constrained by critical time windows or competing components. *Exp Brain Res* 177:440–46.

DWEA. 2003. *Overvågningsrapport 2003*. The Danish Working Environment Authority. http://www.at.dk/Tal/sw13757.asp.

Eurofond. 2007. *Fourth European Working Conditions Survey*. European Foundation for the Improvement of the Living and Working Conditions. http://www.eurofound.europa.eu/pubdocs/2006/98/en/2/ef0698en.pdf

Frost, P., Bonde, JPE., Mikkelsen, S. et al. 2002. Risk of shoulder tendinitis in relation to shoulder loads in monotonous repetitive work. *Am J Ind Med* 41:11–8.

Grande, G., Cafarelli, E. 2003. Ia Afferent input alters the recruitment thresholds and firing rates of single human motor units. *Exp Brain Res* 150:449–57.

Hägg, G. 2000. Human muscle fibre abnormalities related to occupational load. *Euro J Appl Physiol* 83:159–65.

ISO 11228-3. 2007. *Ergonomics—Manual handling—Part 3: Handling of low loads at high frequency.* Geneva: International Organization for Standardization.

Kurppa, K., Viikari-Juntura, E., Kuosma, E., Huuskonen, M., Kivi, P. 1991. Incidence of tenosynovitis and epicondylitis in a meat processing factory. *Scand J Work Environ Health* 17:32–7.

Madeleine, P., Lundager, B., Voigt, M., Arendt-Nielsen, L. 2003. Standardized low-load repetitive work: Evidence of different motor control strategies between experienced workers and a reference group. *Appl Ergon* 34:533–42.

Mathiassen, SE. 1993. The influence of exercise/rest schedule on the physiological and psychophysical response to isometric shoulder–neck exercise. *Euro J Appl Physiol Occup Physiol* 67:528–39.

McGorry, RW., Dowd, PC., Dempsey, PG. 2003. Cutting moments and grip forces in meat cutting operations and the effect of knife sharpness. *Appl Ergon* 34:375–82.

Nordander, C., Ohlsson, K., Balogh, I. et al. 1999. Fish processing work: The impact of two sex dependent exposure profiles on musculoskeletal health. *Occup Environ Med* 56:256–64.

Pedersen, J., Ljubisavljevic, M., Bergenheim, M., Johansson, H. 1998. Alterations in information transmission in ensembles of primary muscle spindle afferents after muscle fatigue in heteronymous muscle. *Neuroscience* 84:953–59.

van Rijn, RM., Huisstede, B., Koes, BW., Burdorf, A. 2009. Associations between work-related factors and the carpal tunnel syndrome—A systematic review. *Scand J Work Environ Health* 35:19–36.

SWEA. 1998. *Ergonomics for the Prevention of Musculoskeletal Disorders.* Statute Book (AFS) 1998:1 Swedish Work Environment Authority. http://www.av.se/dokument/inenglish/legislations/eng9801.pdf

SWEA. 2005. *Occupational Accidents and Work-Related Diseases.* Swedish Work Environment Authority. http://www.av.se/dokument/statistik/officiell_stat/ARBMIL2005.pdf

SWEA. 2008. *Work-Related Disorders 2008.* Swedish Work Environment Authority. http://www.av.se/dokument/statistik/officiell_stat/ARBORS2008.pdf

Wolpert, DM., Ghahramani, Z., Jordan, MI. 1995. An internal model for sensorimotor integration. *Science* 269:1880–82.

FURTHER READING

Buckle, P., Devereux, J. 1999. *Work-Related Neck and Upper Limb Musculoskeletal Disorders.* Bilbao: European Agency for Safety and Health at Work.

David, GC. 2005. Ergonomic methods for assessing exposure to risk factors for work-related musculoskeletal disorders. *Occup Med* 55:190–99.

Johansson, H., Windhorst, U., Djupsjöbacka, M., Passatore, M. 2003. *Chronic Work-Related Myalgia. Neuromuscular Mechanisms Behind Work-Related Chronic Muscle Pain Syndrome.* Gävle: Gävle University Press.

van Rijn, RM., Huisstede, B., Koes, BW., Burdorf, A. 2010. Associations between work-related factors and specific disorders of the shoulder—A systematic review. *Scand J Work Environ Health* 36:189–201.

Schmidt, AA., Lee, T. 2005. *Motor Control and Learning. A Behavioral Emphasis.* 4th revised edition. Leeds: Human Kinetics Publishers.

Sluka, AA. 2009. *Mechanisms and Management of Pain for the Physical Therapist.* Seattle. IASP Press.

Visser, B., van Dieen, JH. 2006. Pathophysiology of upper extremity muscle disorder. *J Electromyogr Kinesiol* 16:1–16.

6 Prolonged, Low-Intensity, Sedentary Work

Allan Toomingas

Photo: Lars Erik Byström

CONTENTS

Suzanne is 47 and has worked full time for some years in the customer service section of an insurance company. She has contact with the company's clients by telephone, e-mail, and sometimes by text message. All the details of the company's products and their clients are on computers. She therefore uses her computer to retrieve information and enter new details into client accounts. A new task introduced a year ago is that, if there are gaps between incoming client calls, she has to ring up potential new clients to offer them the company's services. Altogether, she normally finds time for between 75 and 100 customer calls on each working day. Employees have a fixed monthly salary with a bonus for the number of new clients they recruit.

She gets on well with her colleagues and likes providing a service to clients, but is no "sales type" and feels uneasy at the sales calls she is forced to make.

For many years, Suzanne has felt stiffness in her neck and shoulders. For 6 months she has been troubled almost daily by headaches and a constant ache between her shoulder blades. Sometimes it is difficult to see the screen clearly, and it is a strain to look for different lines in the menu and small text boxes in the various programmes that she has to have open on the screen at the same time.

6.1 FOCUS AND DELIMITATION

This chapter deals with jobs where the physical load is low as regards energy metabolism and muscle force. This is in contrast to Chapters 2 and 3 where it is high. The work is instead marked by being sedentary, with prolonged, low-intensity physical load. The mental load may be high. This chapter does not, however, go into greater detail about the origins of mental load. On this, see Chapter 7.

This chapter, among other things, takes up the following issues:

1. What happens in the muscles after working for a long time in the same work posture?
2. Why can pain result from low-intensity work, for example, when using a computer mouse?
3. Why is there pain in the neck and shoulders when working with one's hands?
4. How does the body react to a low level of activity for long periods?
5. How do we design healthy work if that work is prolonged, low intensity and sedentary, for example, where computers are used for a large part of the day?

6.2 PREVALENCE OF PROLONGED, LOW-INTENSITY, SEDENTARY WORK

There is a great deal that indicates that prolonged, low-intensity, sedentary work has become much more common in working life over recent decades. This is in part a result of the transformation of the economy from the production of goods to the production of knowledge and services. A great deal of these and other jobs is carried out while sitting still, for example, at a checkout, in the driver's seat, or at a computer. Approximately 1/4 of the people working in Sweden state that they work sitting still for more than 2 h at a stretch everyday [SWEA 2010]. Approximately 44% of those people working in Stockholm County in 2006 stated that they had sedentary work. An Australian study from 2005 reported from a community sample that the average self-rated occupational seated time was 4.2 h/day among professionals, 3.5 among white-collar workers, and 2.3 h among blue-collar workers [Mummery et al. 2005].

Studies of objectively measured physical activity in population samples from Australia and the United States report that adults of working age spend 7.3–8.4 h/day at a sedentary activity level, that is, about 50% of their waking time [Healy et al. 2007; Matthews et al. 2008]. One such sedentary occupational setting is computer work. According to the European Working Survey in 2010 among the European Union (EU) countries, 52% of the total questioned used computers for 25% of their working hours or more [Eurofound 2010]. About 29% use them all, or almost all, of the time. The use is somewhat higher among women, but more or less equal among different age groups. The trend towards an increasing use of computers at work can be traced back over the last 25 years in Sweden (Figure 6.1). It can be estimated that roughly a third of all working hours in Sweden in 2009 were spent in front of a computer. Industries in which computers are used a great deal are, for example, customer service work (call centres), offices, education, communication and media work, and the world of banking and finance.

In the production of goods, the work is also often prolonged, low-intensity manual work, for example, in assembly work involving light electrical components. Many processes within the production of goods are automated and computer controlled. The human task is now that of monitoring processes sitting at a control panel or in front of a monitor.

Childhood and school years are also spent sitting still for long periods. Leisure time, too, among young people and adults is often devoted to sedentary activities, for example, TV, computer games, or surfing the internet. Travel to and from work or school often involves sitting still in a private car or on public transport.

6.3 WHAT CHARACTERIZES PROLONGED, LOW-INTENSITY, SEDENTARY WORK?

Sedentary work is often characterized by workers sitting for the greater part of the working day and being active with their hands. Their hands are holding or are active with materials or equipment, for example, a keyboard and computer mouse. Force development in the active muscles is low, in typical cases just a few percent of maximum voluntary strength (MVC), but is exerted almost without a break for long periods. The variation in force development is small, as is the variation in work postures and work movements. This prolonged, low-intensity muscle work is often described as "static" in the literature on ergonomics, even if this is not correct from a strictly physiological viewpoint (see Section 6.6).

6.4 MUSCLES AT WORK

6.4.1 Muscles Perform Work

The muscles are one of the few structures in the body that can exert physical force and in this way bring about movement. Muscles can be found in many places in the body and can vary in size, form, and properties. Usually, we think about some of the ~600 muscles producing movement of the neck, arms, trunk, and legs

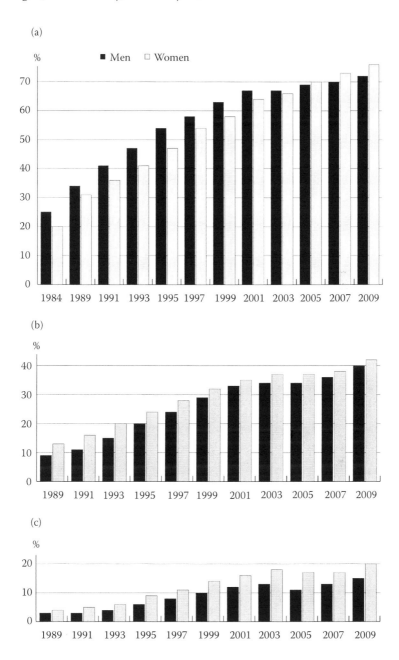

FIGURE 6.1 The proportion of men and women aged 16–64 working with computers: (a) for some part of the working day; (b) at least half of the working day, and; (c) almost all the time. Development during the period 1984–2009. (Modified from SWEA. 2010. *The Work Environment 2009*. Stockholm: Swedish Work Environment Authority, pp. 58–69. http://www.av.se/dokument/statistik/officiell_stat/ARBMIL2009.pdf).

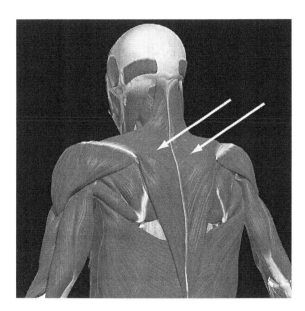

FIGURE 6.2 Superficial muscles of the musculoskeletal system of the upper body. (Arrows: trapezius muscles.) (Photo and copyright: Primal Pictures Ltd. http://www.primalpictures. com. With permission.)

(Figure 6.2). There are, however, muscles also in the heart, in the airways, and in the walls of the intestines and of blood vessels.

Muscles in the body work by exerting force for a certain period. The muscle work in the musculoskeletal system may involve, for example, supporting the body's own weight or moving the body. The work can also involve lifting heavy loads, manipulating various tools, operating controls, or manoeuvring a computer mouse. In this last case, the force required is extremely small, but it is often prolonged.

6.4.2 THE STRUCTURE OF MUSCLES

The muscles of the musculoskeletal system are constructed from densely packed long *muscle cells*. The cells are "wrapped up" in a surrounding membrane, and the ends of the long cells morph into a strong connective tissue which anchors both ends of the muscle to some part of the skeleton, for example, the bones of the forearm. In certain muscles, this connective tissue forms a long tendon running up to a point of attachment on the skeleton, for example, one of the finger bones. The muscle attachments may also be spread across a large area of several parts of the skeleton, such as the source of the trapezius muscle, which runs down the whole way from the back of the skull along the entire length of the cervical and thoracic spine (Figure 6.2). The force that the muscles exert is transferred via the attachment points to the skeleton, which can then move in relation to each other. Different movements are made depending on which muscles are active.

A single muscle cell can be several centimeters long, but is only 0.03–0.1 mm in diameter (Figure 6.3 and Fact Box 6.1). Muscle cells cannot usually divide after

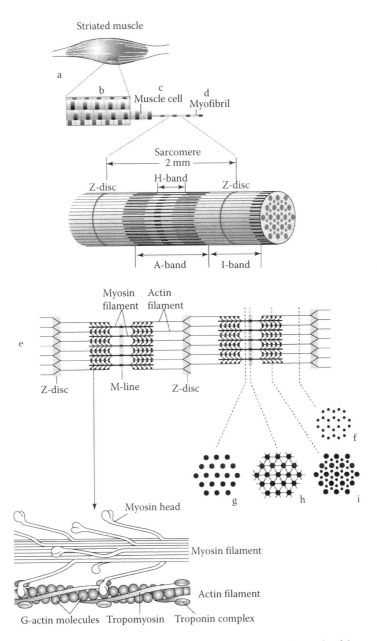

FIGURE 6.3 Building up the muscles of the musculoskeletal system, (a) muscle of the upper arm with magnification of some individual muscle cells (b); (c) individual muscle cell; (d) myofibril where the striation is visible and where the various visible bands I, A, H, and Z are indicated; (e) schematic picture of some sarcomeres (the contractile unit between the Z disks); (f–i) cross section through various parts of the sarcomeres. (Below) Detailed picture of actin and myosin filaments with myosin and troponin protuberances assisting in the sliding of the two filaments against each other. (Modified from Åstrand, PO. et al. 2003. *Textbook of Work Physiology. Physiological Bases of Exercise.* 4th ed. Champaign: Human Kinetics. Illustration: Lena Lyons.)

FACT BOX 6.1

The muscle cell is covered by a cell membrane. There is an electrical potential difference between the outside and inside of the cell membrane, which is called the *membrane potential*. Just as in the nerve cell, this membrane potential comes about through potassium ions "being pumped" into, and sodium ions out of, the cell membrane. The surface of the muscle cell has indentations in the cell membrane at regular intervals which branch out as a network inside the cell, known as T-tubuli. Inside the cell and in close association with the cell membrane around the T-tubuli, there is the sarcoplasmatic reticulum (reticulum = net) which in turn forms a dense network around the myofibrils.

birth, other than in special cases after injury. The number of muscle cells is, therefore, largely constant, and during old age diminishes (see Section 6.8). With exercise, the number of muscle cells does not increase, but the individual cells increase in diameter, so that the entire muscle increases in thickness.

Every muscle cell consists of a bundle of *myofibrils* that are each ~0.001–0.003 mm in diameter ("myo" is the Latin word for muscle) and in their turn consist of long chains of proteins arranged according to a characteristic striated pattern visible through a microscope (Figure 6.3). Two of these oblong proteins, *actin* and *myosin*, partially overlap with each other. The actin and myosin molecules can slip against each other. This movement demands energy. The energy exists in the form of ATP molecules produced in the muscle cell's numerous mitochondria (see Chapter 2, Section 2.4).

6.4.3   TYPES OF MUSCLE CELL

The individual muscle cells in a muscle vary with regard to how quickly they increase their force development and how quickly they tire in continuous activity. One type of muscle cell, known as Type I, provides slow but constant force development with prolonged stimulation. This type works primarily with *aerobic* metabolism (see Chapter 2, Section 2.4). Type I cells are found to a great extent in the muscles of the trunk and legs which hold up the body, the so-called postural muscles. Another type, known as Type IIX, provides rapid force development which, however, rapidly fades away. These cells work primarily with *anaerobic* metabolism (see Chapter 2, Section 2.4). Type IIX can be found to a great extent, for example, in the arm muscles. An intermediate Type IIA has also been described. Different muscles have different proportions of these cell types. Different individuals differ in their proportions, partly because of genetic differences and partly on the basis of how the muscles are used. The proportion of Type I and Type II, respectively, can, for example, be changed through training. Endurance training increases the proportion of Type I, and strength training or sprint training increases the proportion of Type II.

6.4.4 The Innervation of Muscles

Muscle activity is regulated by nerve signals. From the spinal cord a nerve runs together with a blood vessel to the middle of the muscle belly. The nerve consists of many different nerve fibres (neurons). These convey signals from the spinal column and brain out to the muscles activating the muscle for contraction (called α-motor neurons). Other neurons convey signals from the muscles to the spinal cord and brain, such as pain signals. Special neurons run to and from the muscle spindles (see Fact Box 6.2).

FACT BOX 6.2

In the muscles there are receptors that are sensitive to how stretched or contracted the muscle is, known as muscle spindles. In the muscle attachment or tendon from the muscle there are other receptors that are sensitive to the muscle's traction force, called the Golgi tendon organs. In addition, there are receptors in the skin, in the joints and in the joint capsules that signal the position of the joints. Receptors that report pain are numerous in the connective tissue between the muscle cells, around the muscle and in the muscle attachments.

Together, the different receptors and the neurons cooperate in producing functional and coordinated movements and force development, for example, when we are working.

Groups of muscle cells are functionally joined to what are called motor units. A motor unit is linked to one and the same α-motor neuron. This means that all muscle cells in the motor unit receive the same signal for muscle activity, and will therefore be activated simultaneously and with the same force. The cells in a motor unit can be dispersed between other motor units, but are usually to be found in an area which is about 5 mm in cross section. Motor units may consist of anywhere from a few up to 1000 muscle cells. In muscles where motor ability and precision are important, for example, in the muscles of the eye and face, the number of cells involved is low. In muscles that primarily exert great force, for example, in the torso and legs, the number is high.

6.4.5 How Muscles Work

A nerve signal from the α-motor neuron triggers *depolarization* of the membrane potential along the entire surface of all muscle cells in the motor unit. The depolarization spreads into the muscle cells through the T-tubuli (see Fact Box 6.1). Depolarization occurs when so-called ion canals open up for potassium and sodium ions that then stream out of and into the cell. After about a millisecond, the ion canals close and the membrane potential is restored, in that potassium and sodium ions are pumped in respectively out again through an active energy-intensive process. This momentary change in the membrane potential is called the *action potential*.

T-tubuli and the sarcoplasmatic reticulum are a way of rapidly disseminating the simultaneous activation of the myofibrils throughout the muscle cell. In the contact between T-tubuli and the sarcoplasmatic reticulum, the action potential triggers the liberation of calcium ions to the cell fluid around the myofibrils. The increase in calcium content leads to the actin and myosin molecules sliding together a little. The calcium in the cell fluid quickly reverts to its state of rest again; however, in that the calcium ions are reabsorbed into the sarcoplasmatic reticulum through an energy-intensive pumping mechanism. The odd nerve impulse in this way leads to a short, invisible or imperceptible twitch in the muscle. With repeated dense impulses from the α-motor neuron, the myofibrils do not have time to return to their state of rest, but the slippage between actin and myosin molecules continues, so that the overall result is a noticeable shortening and force development in all the muscle cells of the motor unit.

6.4.6   REGULATING MUSCLE FORCE

The force that can be exerted at a particular movement or operation can be regulated in several different ways. The force in a particular motor unit can be amplified by increasing the *frequency* of *action potentials* in the α-motor neuron. The strength can be further increased by *recruiting more motor units.* By recruiting strong motor units, that is to say, those with many cells, the force can be further increased. When a muscle activity begins, motor units with a low number of cells are usually recruited, that is to say, with low maximum force. These are often of Type I. Units recruited for considerable forces usually have numerous cells and therefore contribute greater force. These are frequently of Type II. It is difficult to carry out work demanding precision with simultaneous high force development, as the motor units recruited for high force do not allow precise adjustment of the force. A further way of increasing the force is to *connect up several interacting muscles* (*agonists*) and disconnect the counteracting muscles (*antagonists*).

Force development is at its maximum when the actin and myosin units overlap maximally beside each other, which happens when a muscle is approximately half-contracted. The more the muscle is drawn out or contracted beyond the middle position, the more the actin and myosin units have slipped apart from each other, and the less force the muscle can then produce. The position of the joint is also important, as the muscles are often attached to the skeleton some way from the particular axis, which allows the muscle to use the lever effect that then arises (see also Chapter 3, Section 3.5 and Chapter 4, Section 4.4). Maximum leverage is achieved in the joint position where the lever is at maximum length, for example, angling the elbow joint at ~90° for the biceps muscle in the upper arm. The *orientation of the muscle cells* inside the muscle is also important. Maximum force can be developed in the muscles where all the cells run parallel to the direction of force development. The more the cells are angled away from this direction, the more force is lost.

The *rapidity* of the muscle contraction is also significant for force development. The higher the contraction speed, lesser is the force that can be developed. Maximum force can be achieved when the muscle contracts during stretching, known as eccentric contraction (see Section 6.6). The maximum power generation of the muscle, that is to say, the energy development per unit of time—the product of force and velocity—is, however, normally at about one-third of the maximum speed of contraction.

In the long term, it is possible to increase the maximum muscle force by *training* muscle capacity. Initially, exercise improves the coordination between the different motor units and agonists–antagonists, so that they work more effectively on the precise activity that is being trained. In the longer term, training also increases the number of myofibrils in the muscle cell, so that the individual cells become larger (but not more in number). The cross-sectional area of muscle increases, which is possible to see in, for example, bodybuilders.

6.4.7 Blood Supply to the Muscles

A muscle that is working hard is the tissue which, together with the central nervous system, has the highest energy metabolism per unit of weight in the body. If the muscle works hard for several minutes, the energy metabolism may increase more than 50 times compared with the state of rest. This requires a corresponding flexibility in blood supply, so that the supply of oxygen and energy substrate as well as the removal, for example, of carbon dioxide, corresponds to the need. The blood flow through the muscles can increase from 3–5 mL/s per 100 mL of muscle mass up to 100 mL/s or more. The total energy metabolism of the body muscles may increase from 20 up to 2000 W. In hard physical work, the muscles can take up to ~80% of all blood flow from the heart, compared with 15–20% at rest (see Chapter 2, Figure 2.9).

Muscles are, in consequence, well provided with blood vessels with dense branches (capillaries) between the muscle cells. Cells of Type I have a greater capillary density than Type II. Regulation of the blood flow to the muscle is carried out primarily by the muscles in the walls of the blood vessel contracting so that they cut off the blood flow, or relaxing and thereby increasing the blood flow. Regulation is governed by the sympathetic nervous system. A somewhat constricted flow occurs when at rest. During muscle work, a general increase in the activity of the sympathetic nervous system takes place, which initially provides further general vascular contraction in the body's muscles. Blood flow in the working muscle increases, however, through the local influence of heat development in the muscle, oxygen consumption in the blood, and various metabolites excreted locally in the muscle when it is working, for example, lactic acid, low pH, carbon dioxide, and potassium. In this way, the blood is directed primarily to those muscles that are active.

When the muscle is activated, the pressure inside it increases (*intramuscular pressure*). This increase in pressure occurring inside the muscle can easily be so great that it prevents blood flow to the muscle. In dynamic muscle work (see Section 6.6) muscle contractions alternate with muscle relaxation. In the relaxation phase, the intramuscular pressure drops and the blood flow returns. Initially, a compensatory increase in the blood flow through the muscle occurs, so that the balance is restored as regards, for example, oxygen, carbon dioxide, and pH value. In static muscle work, however, (see Section 6.6), the diminished muscle blood flow can lead to a rapid development of fatigue.

The blood vessels transporting the blood away from the muscle, the veins, are provided with valves inside the muscle so that when the muscle is activated and contracts, the blood is pushed in the right direction towards the heart. Dynamic muscle activity can therefore help blood circulation.

6.5   ARMS AND HANDS ARE FLEXIBLE BUT REQUIRE STABILIZATION

One of the subtleties of the human body is that it is so flexible and agile. This applies particularly to the arms, which through the shoulder blades, shoulder, elbow, and wrist joints are able to provide the hands with almost unlimited opportunities of adopting different positions. This can be used in tasks requiring the hands to work in special positions and with special movements, for example, handicraft, care work, writing with a pen or a keyboard, or manoeuvring a computer mouse. Mobility comes at a price however, namely the price of stability. The hand has to be able to exert force on what it is handling or wishes to affect. In its working position, the hand then has to be stable, and not give way to the counteraction of the object being handled. In order for the hand to be stable in its working position, it is necessary for the whole arm and its attachment to the thorax and spinal column to be stable. If considerable forces are required at work, then the whole body has to be stabilized, and also stand firmly against the floor so as not be moved by the exertion of force.

Those counterforces that arise, for example, when we move a computer mouse, have to be counterbalanced by corresponding muscle activity in the entire system from the neck and out to the hand. In rapid movements, for example, keyboard work, the system does not have time to counterbalance the counterforces in detail. Stability then has to be achieved by generally making the joints extra stiff through activity in the agonists and antagonists (see Section 6.4.6) [Johansson et al. 2003, pp. 83–94]. Stabilization and counterbalancing occur unconsciously and continue as long as the arm and hand are being used, and presumably as long as there is an intention, conscious or unconscious, to use them. Stabilization must exist before the arm and hand begin to be used. This arm–hand stabilizing muscle activity involves muscles all the way from the neck out to the hand. In order to keep the spinal column in the desired position, stabilization also needs to occur on the left-hand side of the neck and shoulder, even if only the right arm and hand are being used. In a corresponding way, there has to be stabilization and coordination between the focusing of the eye and movements in the neck, shoulders, and arms when work requiring hand–eye coordination is being done (see Fact Box 6.3). Generally speaking, we can assume that work requiring precision makes greater demands on the stability of joints than work with lower demands.

In the stabilization of joints and the coordination of movements, muscles are activated that entirely or in part assist or counteract each other's forces and movements, *agonists* and *antagonists,* respectively. A typical example is the wrist, stabilized by agonists and antagonists on each side of the forearm which attempt to bend and stretch the wrist, respectively. Work demanding great precision probably makes great demands on coordination between all the muscles involved. A good

FACT BOX 6.3

HAND–EYE COORDINATION

In the majority of tasks at work and in other contexts where the hands are used, we are dependent on being able to see what we are doing, where objects and tools are located, what they look like and how they should be grasped, moved, and positioned. A typical example of this is manoeuvring a computer mouse, so that the cursor ends up in the right place on the screen. This collaboration requires advanced coordination between the control of the muscles and different sensory organs, such as, eyesight, balance, joint senses, and muscle spindles. The eye has to cooperate with the musculoskeletal system, both with regard to the eye's own muscles which govern its movements, and the muscles of the neck and shoulders, which adjust the position of the head and shoulders, as well as the muscles of the arms and hands, which govern the position of the hand and movements. If coordination between the hand and the eye is impaired, problems may arise—not merely in the form of a poor work outcome, but also in the form of discomfort, dizziness, headaches, and tense and aching muscles of the neck and shoulders.

working technique means that one has a good balance between the activity of agonists and antagonists at an optimal level so that the muscles do not become fatigued too quickly. Fatigued muscles act with poor precision. Factors such as pain and stress may further impair the coordination between agonists and antagonists, which leads to increased activity on their part to achieve the stability and coordination that the work requires.

Stabilization and coordination in intensive manual work therefore requires muscle activity in the neck, shoulders and arms which goes on for as long as the hands need to be used, but which in itself does not lead to any movement. This stabilizing muscle activity is added to the activity needed to carry out the task itself, for example, to manoeuvre the computer mouse.

Knowledge of stabilizing and coordinating muscle activity is, however, as yet deficient.

6.6   STATIC LOAD ON THE MUSCULOSKELETAL SYSTEM

Muscle activity can be carried out dynamically or statically. In *dynamic* activity, the muscle is shortened or lengthened. We can see that the arm, leg, or body part moves, and we understand that the muscle is working. Leg muscles carry out dynamic activity when we walk, the trunk muscles when we bend over and turn around, and the arm muscles when we are hammering or typing on a keyboard. Dynamic muscle activity leading to a shortening of the muscle is called *concentric* activity. For example, the muscles in the back work concentrically when we stretch upwards from having been bending forward. If the muscle exerts force at the same time as it is being extended, this is called *eccentric* activity, for example, when the muscles in the back resist when we bend forward.

In *static* activity, the muscle develops force without changing its length, that is, without achieving any movement. It works *isometrically*. The lack of movement means that it is difficult to see that the muscle is actually active, for example, when the head is held leaning forward, or the arm is held out. In most physiological studies of isometric activity the muscle force development is kept constant in a so-called *isotonic* contraction. Static activity is therefore for a physiologist often both isometric and isotonic. Muscle work of this kind quickly leads to exhaustion. In ~50% of maximum voluntary muscle contraction (MVC), endurance is ~1 min. A lack of blood supply because of intramuscular pressure is a probable explanation for the fact that the force terminates. Lower load levels increase endurance, but even at 5–10% of MVC, there are clear signs of fatigue within 1 h. In that case, it occurs presumably as a result of a shift in the chemical environment in the muscle cells. Old ergonomic recommendations about 15% of maximum voluntary contraction as an upper limit for the load level in prolonged work therefore conflict with today's knowledge.

Within working life and ergonomics the concept of static muscle activity has become fuzzier. "Static" muscle activity is there used to describe circumstances in which the muscle is active for a long period without any major change in either force or position, and with few or no rest periods. Slow and/or minor changes in force and position can occur, which are incompatible with the strict physiological definition above. Such low-intensity activity can continue for a long time (minutes or hours) and initially gives no signals of fatigue or discomfort. The ergonomic significance of "static" muscle work has been given wide circulation, as the lack of muscle rest and inadequate recovery are regarded as the probable cause of many of the disorders that can arise in prolonged low-intensity muscle work (see Section 6.7). "Static" muscle work of this kind in the neck and shoulders can often be found as the basic component in jobs where the hands are busy with tools or materials. It may, for example, be a question of assembly work in the manufacturing industry, sewing machine work at a garment manufacturer, work at the checkout in a supermarket, or at a computer. During computer work, it has been seen that the trapezius muscle works corresponding to 1–10% of MVC for large parts of the work period. Similar levels have been measured from the muscles of the forearm.

Prolonged "static" muscle work without interruption may lead to an adverse strain on the muscles, but also on the muscle attachments, tendons, connective tissue, and joints, if they remain in the same position for a long period.

Prolonged "static" load on the neck, shoulders, and arms can arise for different reasons in a job in which the hands are used [Johansson et al. 2003, pp. 5–46]. One reason may be the need to stabilize the neck, arms, and shoulders in manual work as explained above. If the arms and hands are lifted against gravity, neck and shoulder muscles have to be activated for purely biomechanical reasons. The load on the neck and shoulders could be decreased by leaning the arms and hands, for example, against a tabletop. Load can also occur from maintaining an awkward work posture, for example, by twisting the neck. Precision demands seem to increase the load further. Finally, "static" load can arise for reasons that are not directly motivated by how the work task is to be carried out. It may, for example, be a question of unconsciously shrugging the shoulders in tense or stressful situations. The stress may be linked to time pressures at work, difficulties or conflicts, or other psychological or social reasons. Factors that do not have to do with work at all, for example, worry about one's own or one's family's health, financial worries, or family conflicts, may possibly lead to similar muscle tensions (see Chapter 7, Section 7.11). Pain or other discomfort is in itself stressful, and may lead to increased muscle activity, and therefore in the longer term to even more pain in a vicious circle. Noise disturbance, for example, from ventilation equipment, or colds and draughts can also lead to tense musculature.

Prolonged "static" loads thus arise as a result of factors both within and outside work. There are presumably major differences between people concerning the inclination to use their muscles uninterruptedly and invariably "statically," even when it is a question of carrying out the same task. These differences result both from the choice of working technique, which can be influenced by instruction, and unconscious patterns of motor control (see Chapter 5, Section 5.4).

Prolonged "static" muscle work occurs presumably more rarely during leisure time than at work. Leisure time activities are usually more physically varied, and contain more dynamic muscle work. There are of course exceptions, for example, in computer and television games and the like. During our leisure hours, we are also more at liberty to take breaks and discontinue jobs that may produce strain on the muscles. Nor is there perhaps the same state of stress which is often the case in working life.

6.7   PROBLEMS WITH WORK INVOLVING PROLONGED, LOW-INTENSITY STATIC LOAD

6.7.1   SYMPTOMS

Neck pain is common in the general working population in most countries for which there is reliable information. Typically 30–50% of the workforce are affected on an annual basis, and 10–15% report that it interferes with their daily activities [Côte et al. 2008]. Office and computer workers are found to have the highest incidence of neck disorders. Other high-risk professionals are dentists and other medical staff. Disorders usually include a continuous ache or pain arising in certain body positions or movements. The disorders can be localized, for example, in the neck or between the shoulder blades. In many other cases they are, however, diffuse and difficult to locate. Not unusually, pain is located to other, often more peripheral, parts of the body (so-called referred pain), for example, in the shoulders and arms, or as headaches. Disorders may start as a diffuse feeling of fatigue, stiffness, or tension. In typical cases, the disorders become increasingly intensive, more widespread, and more prolonged. The transition to aches and pains is insidious. Many people also feel pain or tenderness when the doctor palpates the neck and shoulder area, for example, the trapezius muscle. The disorders often affect the neck and shoulder in both halves of the body, even if we use our right hands the most. The reason for this is presumably that muscles in both halves of the body have to be active in order to stabilize the neck and shoulder area (see Section 6.5). Characteristic of the disorders is that they may vary in localization and intensity. Periods of disorder can be triggered by loads at work or life in general. In extreme cases, the disorders lead to prolonged periods of sickness off work and disability pension. Alternative jobs without loads that trigger problems in the neck and upper extremities may be difficult to find.

6.7.2 DAMAGED STRUCTURES

Precisely which structures and tissues are affected in such disorders is in many cases unclear. Medical examinations rarely find any damage, and do not result in a specific diagnosis. We call cases of this kind *non-specific* and we make a symptomatic diagnosis, for example, cervicalgia (= neck pain). In many cases the disorders can be localized to muscles, for example, the trapezius, and are then diagnosed as "neck myalgia" (myalgia = muscle pain). The localization of non-specific disorders is very similar to what we see in those disorders triggered by muscles, which is why it may be asked whether the two types are not closely related to each other (Figure 6.4). Deeper-lying muscles and other structures are more difficult to investigate, which means that their contribution to the disease panorama is less well known.

Microscopic studies of painful muscles have shown various types of damage in Type I cells in the trapezius and forearm [Hägg 2000; Johansson et al. 2003, pp. 95–109]. It has also been found that the muscle cells have increased in thickness. There has, however, been no corresponding increase in the size or number of blood vessels. That might cause a local oxygen deficiency in these muscles. It has also been possible to observe impairments in the function of the mitochondria in Type I cells, which may lead to impairments in the energy supply to the muscle with ATP. It has, however, not been possible to prove conclusively that the damage has arisen directly

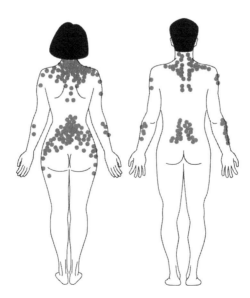

FIGURE 6.4 Distribution across the body of commonly occurring non-specific disorders (left) and muscle disorders (right). Note that these disorders rarely occur at the front of the body or the legs. Illustration: Niklas Hofvander.

from prolonged, low intensity load, or that it is the damaged muscle cells that give rise to the pain.

The muscle attachments and tendons may sometimes be the cause of aches and pains in the neck and upper extremities in individuals with jobs involving "static" load. Usually, tendons and muscle attachments around the shoulder joint are affected, which is called rotator cuff tendinitis. Corresponding problems may arise in the elbow joint, for example tennis elbow, where the muscle attachments of the forearm are affected. Similar disorders can also occur in various tendons around the wrist.

Peripheral nerves that control the muscles or that convey sensory stimulation pass through many narrow passages on their way from the spinal column and to their final destination in the body, for example, in the hand. A prolonged pressure on a tissue—for example, a nerve—impairs its blood supply. Tissues then swell up, which in turn further increases the pressure in a narrow space. When a nerve becomes trapped or exposed to pressure, discomfort arises in the form of numbness, pricking sensations, or pain. This discomfort is often localized to the part of the body which the nerve serves, for example, the hand. Disorders may, however, spread or be diffuse, and difficult to localize in the body. The most common places for nerves in the upper extremities to be trapped is—from the inside and out—the exit of the nerves through the cervical spine; the exit from the neck region out to the arm between the muscles and below the collar bone; and the narrow passages around the elbow joint and in the wrist. One cause of prolonged pressure or the trapping of peripheral nerves may be twisted or bent work postures. A trapped nerve in the wrist, known as carpal tunnel syndrome, is common, which may be caused by prolonged deviated postures or repetitive movements of the wrist, particularly in connection with high grip force. Different individuals vary in sensitivity, presumably depending on different anatomical circumstances in narrow passages, and also depending on hormonal factors that may result in swelling, for example, pregnancy or medical conditions such as diabetes.

Loading of the musculoskeletal system in the lower body seems to be associated with fewer and different problems than those affecting the upper extremities, which may seem to be paradoxical, as, for example, the muscles in the legs constantly help maintain balance when standing up. The difference may be because muscles in the legs are developmentally adapted to constant work. When we are standing, we are also unconsciously varying our centre of gravity, for example, shifting from leg to leg, so that different muscles are active alternately (so-called postural sway). Prolonged standing may, however, result in swelling in the legs and varicose veins (enlarged venous blood vessels). Unsuitable footwear and hard floors can result in disorders of the ligaments and connective tissue of the feet. Prolonged work in a kneeling or squatting position, for example, among floor-layers, may lead to arthritis of the knee joints, presumably because of adverse pressure distribution across the joint cartilage of the knee joint (see Chapter 4, Section 4.4).

6.8   GENDER- AND AGE-RELATED DIFFERENCES

A greater proportion of women than men report disorders of the neck and upper extremities. The reason for this is contentious and presumably multifaceted. One of the explanations given is that women are overrepresented in "static" loaded professions, for example, within administrative work with a great deal of computer use, work at checkouts, or assembly work in industry. But even within the same profession, women can be exposed to greater load than men. This may be because tools are designed for men, and women are forced to adopt more awkward work postures when they use them. One example is the computer keyboard, which in its standard designs is rather wide. Right-handed people like to place the mouse on the right-hand side of the keyboard. For women, this means that their right upper arm is angled outwards more than among men, because generally women's shoulders are narrower. This increases the load on women's muscles, among others the trapezius. Other causes that are frequently mentioned are that women are often burdened with more stress factors, both at work and in the family. It should be added that several studies have found that women are more pain sensitive and more affected by pain. This may also explain part of the difference in reporting disorders between the genders as regards the neck and upper extremities.

Older people lose muscle cells in the motor units, which then become weaker [Åstrand et al. 2003, Chapter 4]. This applies particularly to muscle cells of Type II. The entire muscle becomes thinner (sarcopenia). The number of α-motor neurons also declines. The α-motor neuron from other motor units may then grow out and take over control, which, however, makes for a reduction in precision in the movements. It is well known that speed, coordination, and balance decline, particularly with advanced age. With increasing age, the elastic components of the musculoskeletal system degenerate, for example, in muscles, tendons, connective tissue, and joint capsules. Over the years you get stiffer! Stimulation in the form of well-judged dynamic physical load can compensate for the loss of force and coordination in such age-related changes. It should therefore be particularly important to maintain these functions through compensatory training and avoid monotonous, prolonged periods of sitting still.

6.9 EXPLANATORY MODELS

6.9.1 DIFFERENT MODELS

There is no clear-cut explanation of why prolonged, low-intensity, "static" work leads to disorders and illness in the musculoskeletal system. Various explanatory models describe different pathogenetical mechanisms [Huang et al. 2002; Johansson et al. 2003, pp. 5–46; Visser and van Dieën 2006]. Some describe factors leading to uninterrupted "static" activation of the muscles; others explain how the muscle reacts to such activation. Further theories describe how pain can arise, be sustained, and be experienced. The models do not contradict each other, but need to be combined for us to be able to understand the entire sequence of events between workload and experience of pain. It is probable that there are different mechanisms, the significance of which varies at different stages of the development of the disorder, as well as in different situations and for different people.

Figure 6.5 provides a general model of how some of the most common explanatory factors may be thought to act together in the genesis of disorders involving prolonged, low-intensity, static muscle activity. As described above, work tasks,

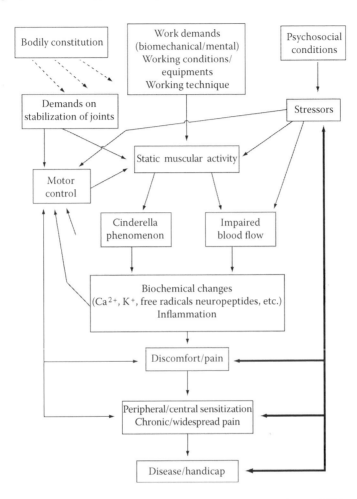

FIGURE 6.5   Model of factors involved in the emergence of disorders and ill health in the muscles with prolonged, low-intensity, static muscle activity. Note the two-way arrows between stressors and pain/illness where stress can influence the emergence of pain and illness, which in turn can give rise to stress. Note also several other possible vicious circles, where the result of static muscle activity by means of the Cinderella phenomenon or impaired blood flow and biochemical changes and pain may increase static muscle activity. Several of these factors impair motor control of muscle activity or give rise to stress. Both of these can in their turn further increase static muscle activity. See the text for more detailed explanations.

equipment, or tools in many cases are such that they give rise to "static" muscle activity. In other cases, also described above, it may be a question of unsuitable working technique without rest and support for the arms and hands. In addition, adverse psychosocial conditions and various stressors may increase muscle tension. The need for stabilization, described at the beginning of this chapter, is a further factor that may contribute to the "static" muscle activity. The significance of motor control and the interaction with psychological stressors in the genesis of, for example, the Cinderella phenomenon and impaired blood circulation, changes in the biochemical environment, inflammation, and pain are discussed below. It should be borne in mind throughout that individual differences exist at many different levels which modify the effects, among others as regards physical constitution/anatomy, motor muscle control, coping with stress and pain, and illness. This may explain why not everyone reacts the same way to similar occupational contexts.

Some of the commonly occurring explanatory models are described below.

6.9.2 THE CINDERELLA MODEL FOR MUSCLE ACTIVATION

The explanatory model that has perhaps attracted the greatest attention is what is known as the *Cinderella model* [Johansson et al. 2003, pp. 127–132]. The background is that there seems to be a system for how motor units are connected (recruited) when a muscle increases its contractive force from zero upward. First Type I cells are recruited, initially weak ones. If the force needs to be increased further, gradually more and more powerful motor units are recruited and more of Type II (Figure 6.6). When the muscle then reduces the contractive force, the motor units are disconnected in the reverse order from that in which they were connected. This means that those motor units connected last will be the first to be disconnected. Those that were connected first will be active for the longest time and will be the last to be disconnected. Just as in the case of Cinderella in the fairytale, these "Cinderella units" will be the first to be woken up to go to work and the last to go to their rest. Unlike heavy muscle work, which provides clear feelings of fatigue and discomfort, prolonged low load does not give any necessary signals that it might be a good idea to take a break or change working technique. The Cinderella units can therefore remain active for a long period. The model thus describes why certain muscle cells might be active for a long period, but provides no explanation as to why this is disadvantageous. This is explained by other models (see Sections 6.5 and 6.9.3 through 6.9.6).

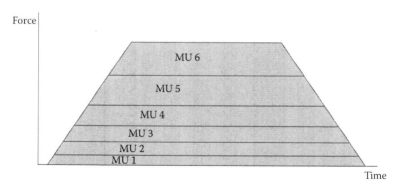

FIGURE 6.6 Illustration of the Cinderella model showing how motor units (MUs) are recruited in a particular order, MU1–MU6, when force development increases in a muscle. When the force then decreases, the motor units are phased out in reverse order so that the units first recruited are phased out last and therefore are active longest. Note that the units recruited last contribute greater force than those recruited first.

The Cinderella model has been verified in several studies where *electromyography* (EMG) signals have been recorded (see Section 6.12.6) from individual motor units in the trapezius or forearm muscles. The Cinderella units have been found to be active for long periods of isometric and isotonic muscle contraction, just as in computer mouse work. The same units have also been active in slow movements in different work postures. Experiments have also shown that the same motor units as those activated in physical work can also be active during mental stress, when no physical work is taking place.

Studies have shown that the Cinderella pattern does not always occur, however, and that it can be varied in different situations. Motor units seem also to be able to work according to a substitution or stand-in principle, where an active unit can be replaced by another with the same function during an ongoing contraction. Some individual studies have found indications that such an exchange between active motor units can be stimulated by the muscle making a powerful contraction, or the opposite, that there is a brief pause.

The Cinderella model emphasizes that it is not enough to reduce the load level. If the level does not decrease to zero, there will be active Cinderella units that in the longer term may be damaged. Healthy work must therefore be based on a variation between activity in different muscles and/or muscle rest (recovery). The substitution principle also opens the door to motor units within the same muscle relaxing alternately, and perhaps finding the necessary recovery, even when the muscle in its entirety continues to work.

One example of the importance of muscle rest may be found in a study of female food production workers who filled chocolate creams in chocolate boxes on a continuous belt [Veiersted et al. 1993]. The ability to relax and rest the shoulder muscles (trapezius) differed between individuals when there was an involuntary break in the flow of chocolate creams on the belt. Measurements showed that some of the women relaxed and allowed their shoulder muscles to rest on such occasions. Others kept their muscles tensed. The study showed that within a few months considerably more of the women who did not relax developed pains in the neck and shoulders. Other studies of computer workers, medical secretaries, and checkout staff in supermarkets have shown that people with muscle pain have fewer episodes of muscle rest of the trapezius than those who are pain free, even if it is impossible to determine whether the lack of muscle rest gives rise to, or is a result of, the pain.

6.9.3   IMPAIRED BLOOD FLOW AS A CAUSE OF PAIN

Another explanatory model that is often put forward as to why constant activation of a muscle might lead to damage and pain is based on the fact that the blood circulation then becomes impaired. When a muscle contracts, the intramuscular pressure increases as described above. In this way, the blood circulation through the muscle worsens. In static muscle work, as opposed to dynamic, no muscle relaxation occurs, which may lead to the intramuscular pressure remaining at a level that prevents blood flow. This may, for example, be the case for the supraspinatus muscle located at the upper edge of the shoulder blade just under the trapezius muscle [Järvholm et al. 1988]. The supraspinatus muscle lifts the upper arm outwards or forward in the shoulder joint. Pressure measurements have shown that, even in a lift of 30° in the shoulder joint, the pressure increase has affected the blood flow. A greater lift increases the pressure. Insufficient flow can lead to oxygen deficiency in the muscle, and in this way to a number of biochemical processes that can produce pain. A well-known process in oxygen deficiency is the formation of lactic acid which results in a lower pH value. In a corresponding way, it has not been possible as clearly to prove reduced blood circulation in the trapezius muscle during low-intensity contractions. Impaired blood circulation has, however, been measured in the trapezius muscles of people with muscle pain [Johansson et al. 2003, pp. 111–115]. Whether this is the cause or effect of the pain is unclear, however. It has not been established whether similar phenomena occur in other muscles. Muscles lying close together and surrounded by bone or taut fascia (connective tissue) do not have much space to expand into. The pressure increase there may become significant even at low levels of contraction. Presumably, there are also individual differences in sensitivity to static load that from this viewpoint are dependent on differences in anatomical circumstances.

Stressors can, by their vasoconstrictive activity in the sympathetic nervous system, also impair blood flow (Figure 6.5).

Another problem is that blood supply to the supraspinatus tendon which comprises a continuation of the supraspinatus muscle out to the upper part of the humerus mainly goes via blood vessels through the muscle. Impairment of the blood circulation may therefore also affect its tendon. This tendon is located in a confined space in a bone channel in the shoulder blade where, moreover, there might be parts with meager blood supply. The tendon therefore has a poor blood supply, even under normal conditions. Tissues affected by oxygen deficiency swell up. The swelling in turn contributes to a further pressure increase, and it is possible then to end up in a vicious circle of impaired blood circulation—oxygen deficiency—swelling—pressure increase as a result. The result can be inflammation of the tendon (rotator cuff tendinitis) with subsequent scar formation and weakening. The confined space in the bone channel can vary between individuals, which may explain why different people have varying sensitivities to static work from this aspect.

The model therefore stresses that impairments in circulation are one explanation for musculoskeletal disorders. There are, however, research findings that contradict the idea that impair blood circulation is the only cause of such disorders.

6.9.4 ALTERED BIOCHEMISTRY AS A CAUSE OF PAIN

Some researchers have emphasized the possibility that the calcium ions released in the muscle cell when it is activated stimulate the release of destructive enzymes which cause intracellular and extracellular tissue damage (Figure 6.5). This, together with possibly increased extracellular potassium levels, may lead to pain stimulation of the nerve endings and release of neuropeptides, for example, bradykinin and prostaglandins, which in turn stimulate the pain receptors. Another explanatory model is based on the fact that the so-called free radicals are formed in connection with the blood flow suddenly increasing in the muscle after having been shut off during muscle contraction. Such free radicals are very reactive, and it is feared that they damage the membranes of the muscle cells, enzymes, and the ion pump. Whether this is a problem at low levels of local static load is controversial. Impairments in the biome-chanical environment lead to different *inflammatory processes* being put into action (Figure 6.5).

6.9.5   Motor Control in Prolonged, Low-Intensity Sedentary Work

The chemical substances formed in a local inflammation may presumably intensify the tendency on the part of the central nervous system to recruit the motor units according to the "Cinderella principle." If that is the case, then a vicious circle devel-ops, in which constant muscle activation leads to chemical changes, which lead to an even greater risk of constant activation (Figure 6.5). The muscle spindles play a cen-tral role in this circle. Release of pain-stimulating substances in the muscle stimu-lates the pain-sensitive neuron, which leads to the spinal column and there affects outgoing nerves to the same and other muscles. A special effect occurs in the so-called γ-motor neuron that regulates the sensitivity of the muscle spindles [Johansson et al. 2003, pp. 291–300]. Other pain stimuli, for example, from joints, also have the same effects. According to the model, the chemical changes reduce the sensitivity of the spindles, and the motor activity becomes "coarser."

These impairments can spread not only to nearby muscles, but also to muscles in the other half of the body. The result is a spread of the impairments in proprioception and coordination of the muscle movements. Impairments in coordination lead to greater activity in the agonists–antagonists, which puts further load on the muscles. A greater need also arises to stabilize the joints, particularly centrally in the body, which also leads to greater muscle activity.

The sensory neurons in the muscles which react to inflammatory substances also affect the activity in the sympathetic nervous system, and in this way the blood circulation, which becomes impaired. Activity in the sympathetic nervous system can also further impair the muscle spindle function [Johansson et al. 2003, pp. 243–273].

6.9.6   Sensitization and Propagation of Pain

Stimulation of free nerve endings (pain receptors), for example, between muscle fibres, causes them to secrete neuropeptides (such as substance P) which stimulate nearby mast cells to secrete, for example, histamine, prostaglandins, and leukotri-enes, or other substances leading to pain and *inflammation* through dilation of the blood vessels, swelling and an accumulation of white blood cells [Johansson et al. 2003, pp. 207–224]. These substances increase the sensitivity of the pain nerves, so that heat, cold, or touch triggers pain (peripheral sensitization) (Figure 6.5). In addi-tion, the area sensitive to pain increases in extent as a result of altered activity in the central nervous system (central sensitization). These phenomena may explain how pain can become more extensive than what is justified by the original injury. The altered activity in the central nervous system may continue even after the original source of the pain has healed. The disorders have then gone from having a local source, which is perhaps quite easy to remedy, to becoming central (in the nervous system) and then being much more difficult to deal with.

Pain is experienced by the individual affected. How it is experienced, its intensity, character, and, above all, its emotional significance—irritating, threatening, or anxiety-inducing—varies between individuals and different situations (Figure 6.5). Circumstances in the individual's environment, particularly those with a positive or negative influence on their emotional state, may affect their experience of pain.

Pain and illness are often stressors in themselves, and can therefore be added to the other stressors affecting activity and blood flow through the muscle (Figure 6.5). Here, too, a vicious circle can develop, where stress leads to static muscle activity and an impairment of muscle coordination and blood circulation with the resulting development of pain and more stress.

6.9.7 OTHER EXPLANATORY MODELS

There are further models that attempt to explain the muscle pain, either as a result of friction between muscle cells [Johansson et al. 2003, pp. 117–126], as a result of hyperventilation, or as a condition similar to migraine with dilated blood vessels. None of these models are today generally accepted.

6.10 SCIENTIFIC EVALUATIONS

The overall picture from research shows that there is *evidence* for a link between prolonged, low-intensity, static work and disorders of the musculoskeletal system. Reviews that have been published during the last 10–15 years have looked at varying aspects of strain and disorders with diverging criteria, and therefore arrive at rather different conclusions. A major review from the United States does not specifically take up "static" load as a risk factor [Bernard 1997]. It does, however, conclude that there is strong evidence for a link between awkward work postures and disorders of the neck and upper extremities. A Dutch review considers that sedentary work for a large part of the day with a static work posture for the neck and upper extremities without breaks constitutes a risk for neck disorders [Sluiter et al. 2001]. Keeping the arm outstretched without support for several minutes repeatedly for most of the day constitutes a risk for disorders in the shoulder and upper arm. A review from the United Kingdom concludes that there is evidence that static loading of the neck and shoulder muscles in combination with repetition and neck flexion increases the risk of neck pain and pathological findings in a medical examination of the neck [Palmer and Smedley 2007]. A recent joint international review from 2008 found evidence that sedentary work is a risk factor for neck pain [Côte et al. 2008].

6.11 SITTING STILL: A HEALTH RISK IN ITSELF

6.11.1 SEDENTARY WORK IS PHYSICALLY LOW-INTENSITY WORK

Prolonged low-intensity work is often of generally of low intensity as regards the demand for energy metabolism (see Chapter 2, Section 2.3). Work is often carried out when sitting still, for example, at a computer. Classic office work required a certain amount of handling of, for example, papers, files, post, and copying. In computer work, many of these functions can be dealt with while seated at a computer. Communication with colleagues at work and others is carried on via the intranet, e-mail, and the internet. Office workers in a typical case walk between 3000 and 4000 steps everyday, compared with 9000–10,000 steps for manual workers.

Older people usually have a more immobile lifestyle and move less than younger people. The number of steps that older people take is usually between 20% and 40% fewer than among young people. Age in itself reduces the body's maximal aerobic capacity and muscle strength. This age effect is added to the negative effect of less physical activity, so that many older people among the workforce do not have sufficient fitness and strength to manage physically more taxing work tasks (see Chapter 2, Section 2.9 and Chapter 3, Section 3.7).

6.11.2 SEDENTARY WORK COULD BE A HEALTH RISK

There are few studies that have proved negative consequences for health from prolonged sedentary work. Most manifestations of ill health from such work may be expected to arise only after many decades, often after the age of 50–60. Carrying out studies in which one follows the working conditions and state of health of groups for decades is very expensive and difficult to do. The effects of activity in working life must also be held separate from the effects during leisure time and other factors, for example, smoking, diet, and social group. For this reason, only a few studies of this kind have been carried out worldwide.

One such study from London was able to show as early as the 1950s that sedentary bus drivers had a higher incidence of cardiovascular disease than the bus conductors who ran up and down the stairs of double-decker buses [Morris et al. 1953]. Postmen ran less risk than staff who worked at the post office. Recent studies have shown that prolonged sitting still in itself, for example, at work, may comprise a risk of serious and potentially life-threatening illnesses, including diabetes, cardiovascular disease, and cancer (see also Section 6.11.4) [Mummery et al. 2005; Moradi et al. 2008; Ekblom-Bak et al. 2010; van Uffelen et al. 2010; Lynch 2010]. This risk seems to arise irrespective of whether one is otherwise physically active, for example, keeping fit during the leisure time. The causal mechanisms are as yet unclear. An adverse influence of sitting still for prolonged periods has been found on lipid metabolism [Katzmarzyk et al. 2009]. Information about how much physical activity is needed to avoid these risks is, however, limited. The conclusion we can draw from these new findings is that sitting still for prolonged periods should be avoided both at work and in leisure time. Proposals for measures remedying this can be found in Section 6.14. Lack of physical activity was rated as one of the most important emerging physical risks related to occupational safety and health by 66 experts invited by the European Agency for Safety and Health at Work [European Agency 2005].

Studies have shown a divided picture of the link between the demands of work on physical activity on the one hand and physical capacity and ill health on the other [Savinainen et al. 2004]. This is paradoxical, as there is clear proof of a positive link between physical capacity, health, and physical activity during leisure time. Possible explanations of this paradox are that physical activity at work is not sufficient to achieve a noticeable training effect, that the time for recovery is insufficient, or that other factors at work, for example, stressors, influence the training effect. It is known that psychosocial stressors, for example, a lack of control at work, increase the risk of cardiovascular disease.

6.11.3 The Continuous Remodeling of the Body

6.11.3.1 A Model

The cells and tissues of the body, for example, muscles, connective tissue, and tendons, do not remain in a steady state after development during puberty. They are constantly replaced by being broken down (*catabolic processes*) and rebuilt (*anabolic processes*) in a remodeling process [Åstrand et al. 2003, Chapters 3, 7, 11]. Figure 6.7 shows a possible model of how physical activity at work or during leisure time (e.g., exercise) provides stimulation to a reinforcement of structures under strain in the musculoskeletal system and cardiovascular system. The lack of load results in a weaker structure. However, sufficient nutrients are necessary for reconstruction, that is to say, energy-rich substrate and building blocks (e.g., amino acids). The must also be *time* for recovery after activity. It can be assumed that the

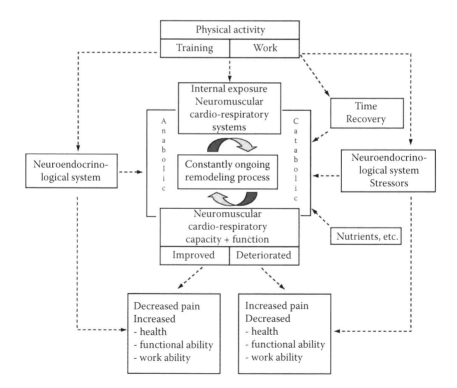

FIGURE 6.7 Illustration of factors contributing to the constant catabolic–anabolic remodeling process in the body. The outcome can be better health, function, and a better ability to work. Or vice versa. See the text for more detailed explanations.

lack of recovery time leads to the opposite effect in the form of accumulating (micro-) injuries. The lack of time for recovery can also be one of the reasons as to why people with high physical load at work (heavy work) often have poor fitness, strength, and endurance in their musculoskeletal systems (see Chapter 3, Section 3.8.2). Anabolic hormones—growth hormone, for example—can stimulate reconstruction. Growth hormone is secreted primarily during sleep. Major sleep disturbances may impair this process. Stress hormones are catabolic, which is why chronic stress may impair reconstruction. The result of an optimally stimulated reconstruction may, according to the model, be better health and an increased capacity to work and function. The result of a lack of stimulation or too high a load may be the opposite—poorer health, and a decreased capacity to work and function. All things in moderation!

Of the body's proteins, about 0.3–0.4% is replaced everyday [Åstrand et al. 2003, Chapter 11]. Roughly speaking, we can say that all the proteins in the body will have been replaced within a year. The rate of this turnover varies a great deal, however, between different tissues. Cells in the gastrointestinal tract, for example, in mucous membranes and body fluids are quickly replaced. Cells in the nervous system, for example, are replaced very slowly, if at all.

6.11.3.2 Muscles

Of the muscle proteins, ~0.1% is replaced everyday [Åstrand et al. 2003, Chapter 11]. Of the contractile muscle proteins, about half are estimated to be replaced over a 1-to-2-week period. Physical activity reduces breakdown (particularly in Type I cells) and stimulates new synthesis (particularly in Type II). This also applies to the heart muscle cells. The strength and quality in the musculoskeletal system, for example, muscles, ligaments, and tendons, decreases with inactivity. After a month's immobilization in a plaster cast, for example, they have fallen by half. It then takes a long period to recover strength.

Physical activity also increases mitochondrial activity in the muscles, and thus the ability of the muscles to use oxygen and nutrients to form energy-rich ATP molecules that are necessary for muscle function. In physical inactivity, the maximal aerobic capacity (fitness) decreases [Åstrand et al. 2003, Chapter 11]. Half of the body's oxidizing enzymes are replaced within a week, and the glycolytic enzymes are replaced within a few days. Aerobic capacity falls by ~25% after 3 weeks of bed rest. See also Chapter 2, Section 2.12 and Chapter 3, Section 3.8.2.

6.11.3.3 Joints

Joint cartilage is not provided with blood vessels, but is dependent on, for example, oxygen and nutrients being supplied from the environment, just as it is on metabolic products being removed. This transport occurs with limited capacity during rest, through so-called passive diffusion and osmosis. A more efficient method of boosting the transport is to dynamically load the joint cartilage. We then achieve a sponge effect in which the joint cartilage alternately absorbs fluid which is then pressed out [Åstrand et al. 2003, Chapter 7]. Inactivity therefore makes for a poorer exchange of nutrients for the joint cartilage. A static compression of the cartilage, for example, because the joint is being held for a long time in a particular position without movement, may be particularly damaging, as it hampers even this weak transport through diffusion. See also Chapter 3, Section 3.8.

6.11.3.4 Skeleton

The skeleton is a living tissue. Cells found in all bone tissue, for example, in the femur, are constantly breaking down the bone structure and others are rebuilding it again. We can provide a rough estimate that after approximately 10 years this replacement corresponds to the weight of the entire skeleton [Åstrand et al. 2003, Chapter 7]. Unfortunately, from the age of about 20 a small amount of bone mineral and bone tissue is being lost, so that the skeleton in time becomes less robust (osteoporosis). This loss increases if the skeleton is not loaded. We should, therefore, load the skeleton for the equivalent of 3 h/day by standing or walking to reduce this loss [Åstrand et al. 2003, Chapter 7]. It is known that in postmenopausal women osteoporosis increases particularly rapidly. It is, therefore, particularly important that in this group harmful inactivity is avoided and favourable loading stimulated.

6.11.3.5 Coordination and Balance

Inactivity also impairs the coordination of muscle movements and balance in the body. The risk increases of faulty manipulation, stumbling, dizziness, and accidents. This, in combination with age-related weaker musculature and skeleton, substantially increases the risk of falls and fractures among older people. The risks of prolonged sitting still are therefore particularly high for older people.

6.11.3.6 Body Fat

Low physical activity leads to low energy metabolism in the body. It is not possible to increase it through mental activity, because the total energy metabolism of the nervous system is relatively unaffected by mental activity. Energy metabolism at rest is ~0.9 MET (For MET see Chapter 2, Section 2.3) [Ainsworth et al. 2000]. The energy metabolism in sedentary computer work is ~1.5 MET, ~2 MET in traditional office work, and 2–3 MET in, for example, sedentary assembly or monitoring work. A 60 kg individual in sedentary work of this kind therefore metabolizes between 720–1440 kcal (3.0–6.0 MJ) over an 8-h shift. In more active professions, the energy metabolism is 4–5 MET, which corresponds to ~1920–2400 kcal (8.0–10.0 MJ) over the same period. A difference of 1000 kcal (4.2 MJ) per working day corresponds to the energy we take on board when we eat ~100 g of pure fat or 250 g of sugar or other carbohydrates. An individual weighing 60 kg working full time, who transfers from an active and moderately strenuous job, for example, as a nurse (4 MET) to low-intensity, sedentary work, for example, providing medical information over the phone (2 MET), may count on a weight rise of up to ~13 kg over one working year if they continue to consume as much food (energy) as in their previous job. In this calculation, we have taken into account the fact that, the heavier one becomes, the more energy is metabolized at a certain level of activity.

There is therefore a risk, particularly in low-activity jobs, that the energy intake through food exceeds what is metabolized during the day, if there is no increase in metabolism during leisure hours. The risk of becoming overweight and obese is clear.

6.11.4 HEALTH RISKS FROM PHYSICAL INACTIVITY

Low physical activity and being overweight/obese may make for a risk of developing *Type II diabetes* and an increase in harmful blood lipids [SNIPH 2010]. It has also been found that there is a link between a lack of physical exercise and a greater risk of *high blood pressure.* Each of these comprises a risk factor in the development of *cardiovascular diseases* such as angina, heart attack, and stroke. Prolonged physical inactivity may therefore have serious complications. Facts indicate that the risk of such health problems is more closely linked to the lack of physical activity than to being overweight. It may therefore be more dangerous to be of normal weight but inactive than overweight but physically active.

In recent decades the indications have been growing of a link between lack of physical activity in the form of exercise and the development of *cancer* [Dept. of Health 2004; Lynch 2010]. This is primarily a question of breast cancer in women and prostate cancer in men. In addition, there is a link to cancer of the gastrointestinal tract. Low physical activity has also been linked to motor impairments in the gut with an increased risk of constipation, haemorrhoids, and diverticulosis.

Many of the health consequences that have been mentioned can also be influenced by different stressors, such as adverse psychosocial conditions, primarily high mental demands, a lack of control over one's own work, and lack of social support from management and colleagues [Palmer and Smedley 2007; Côte et al. 2008]. Sedentary work with poor psychosocial conditions and other stressors may therefore be assumed to be particularly unhealthy.

Apart from the link between physical activity and physical health, in recent times information has also come to light about the link with *mental well-being* [SNIPH 2010]. Physical exercise seems to reduce sensitivity to stress. It may also provide a more positive self-image, and satisfaction with oneself as an individual. The general sense of well-being increases. Exercise has also been proved to have a positive effect on *depression. Sleep disturbances* are another problem area in which physical activity may have beneficial effects. The incidence of psychological disorders of this kind is higher among women and among older people.

All in all, it has been shown that physical activity in the form of physical exercise has a positive effect on a large number of both physical and psychological functions. Positive effects can in most cases be seen against an existing state of ill health, both as directly healing and as enhancing the quality of life in chronic illness. In many cases a preventive effect against the emergence of such ill health can also be seen. Inversely, WHO reports that physical inactivity is among one of the 10 most common causes of death in the developed countries. It also causes a large proportion of the losses of health (Disability Adjusted Life Years—DALY = the number of years lost with full health where account has been taken of how serious the health loss is)— 23% of them linked to cardiovascular disease, 16% to colon cancer, 15% to Type II diabetes, 12% of strokes, and 11% linked to breast cancer.

The link between these widespread diseases and physical activity *at work* is in most cases unclear, but information about these links is growing (see Section 6.11.2). It is unclear as to what proportion of the cases of these illnesses can be ascribed to periods of low-intensity, sedentary work. Most of the illnesses have a multifaceted background, but with a probable influence both from working life and from life in general. A working life becoming more sedentary will scarcely contribute to a positive development of public health, however.

6.12 IS THE JOB LOW INTENSITY, SEDENTARY, OR WITH STATIC LOAD?

There are different methods of assessing general physical activity and specific muscle activity in connection with work: (a) occupational designation; (b) self-reported activity; (c) observation of work; (d) using measuring devices.

6.12.1 OCCUPATIONAL DESIGNATION

A rough way of assessing physical activity and energy needs at work is to use occupational designations, trade designations, or the like. There are tables of energy use for different professions and activities [Ainsworth et al. 2000]. The advantage of this method is that it is very simple and inexpensive. The disadvantage, however, is that it provides a very rough picture, as it does not take into account the specific circumstances of the workplaces and individuals involved. The information can also become outdated and give a distorted picture of the current situation. Information about muscle load also rarely occurs in registers of this kind.

6.12.2 SELF-REPORTED PHYSICAL ACTIVITY AND LEVEL OF EXERTION

Perhaps the most common method of getting some measure of the activity is that the individuals themselves assess the degree of activity or exertion at work. Usually, a questionnaire with scales is used, where it is possible to choose the level of work among specific alternatives, for example, how "heavy" the job has been, how long the activity has continued, and how often. The advantage of this type of assessment is that the individual is being instructed to think about a particular time period, for example, "the past year" or "a typical day." It is also possible to use a questionnaire to obtain information from many individuals at relatively low cost. The disadvantage is that the reliability and validity of the responses may be questioned. The quality of the responses is dependent on the individual remembering correctly and not allowing irrelevant factors to influence their response.

Factors often affecting assessments of activity in intensity levels are the individual's frame of reference and working capacity. If two people have the same job, one who is seldom physically active will presumably assess the incidence of certain physical activity as occurring more often, for example, how often you do "heavy work," than another person who is more used to physical activity. An individual who, for example, is out of shape or lacks muscle strength will also be more laboured in a given physical job and therefore presumably describe the intensity as higher than someone who has good fitness or strength. It is, therefore, often more reliable to ask directly how *strenuous* the individual experiences a particular job as being, and present the response according to a scale, for example, Borg's Scale (Figure 6.8) [Borg 1998]. Exertion is the *internal exposure* that occurs when an individual with a certain capacity is subject to certain

6	No exertion at all
7	
8	Extremely light
9	Very light
10	
11	Light
12	
13	Somewhat hard
14	
15	Hard (heavy)
16	
17	Very hard
18	
19	Extremely hard
20	Maximal exertion

FIGURE 6.8 Borg's rated perceived exertion (RPE) scale 6–20. The scale values 6–20 are set so that, when multiplied by 10, they correspond to the pulse in short-term dynamic work by major muscle groups, for example, on a bicycle ergometer, in a healthy 20–25-year-old person of average fitness.

external exposure, for example, cycling at 100 W or lifting 25 kg. The Borg Scale is constructed in such a way that the endpoints comprise "no exertion" and "maximum experienced or imaginable exertion" respectively, that is to say, a range that is reasonably similar for all individuals. An assessment, for example, in the middle of the scale, therefore means an approximately equivalent experience of exertion for everyone.

6.12.3 OBSERVATIONS OF WORK

Another method that is frequently used to measure physical activity is to observe the individual's activity at work. Trained observers can, with the aid of a stop watch, record the incidence and timing of some predetermined activities. They can, for example, observe the time taken for sedentary and standing work, the time with the hands above shoulder height or the neck twisted. In order to facilitate observations of this kind there are special computer programmes for small palm computers on which one records one's observations and which can also calculate results, for example, the proportion of time in sedentary work. As an alternative to observations on site, one can, using a web camera or video recorder, record sequences of work so as to later register the activity from the recording in peace and quiet. By inspecting the recordings several times, one can register considerably more than one has time for in real time on site. The advantage of observations is that it is possible to avoid the influence of such irrelevant factors as might affect self-assessment (see Section 6.12.2). There is also a broad spectrum of activity parameters to choose from. The disadvantage of observations is that they take a long time and are therefore expensive, and with reasonable resources it is possible to observe for only a limited period of time. The reliability and validity of measurement results depends on the observer having captured sufficiently long and realistic "time windows." Quality is also dependent on having

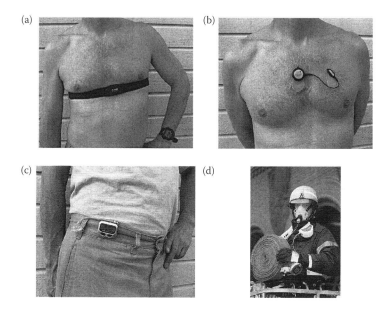

FIGURE 6.9 Personal monitoring equipment: (a) Heart rate monitor ("pulse meter"); (b) heart rate monitor with accelerometer; (c) pedometer; (d) oxygen consumption logger—MetaMax 3B. Photo: (a–c) Martin Toomingas; Photo: (d) CORTEX.

chosen the relevant aspects of physical activity and that these have been observed with high reliability and validity.

6.12.4   MEASURING PHYSICAL ACTIVITY

There is, nowadays, a selection of measurement instruments for recording physical activity which can be carried on the person during a working day or over several days. The basic principle for most of them is that *the body's position and movements vis-à-vis gravity* are recorded. Pedometers are commercially available which can be attached to a belt or the like (Figure 6.9c). Other instruments that measure whether the subject is sitting or standing/walking include *inclinometers* and *accelerometers*. A more or less sophisticated computerized programme software is connected to the various instruments to calculate activity during the recording period. The advantages of instrumental monitoring of body movements is that many irrelevant factors that may affect the assessments are avoided. It is also possible to record over a long period, from days to weeks, and in this way to gain quite a good understanding of the activity. The cost of these instruments varies, but for example, pedometers are fairly cheap. The quality is dependent both on acquiring relevant aspects of physical activity and that the measurements are technically reliable.

6.12.5   MEASURING ENERGY METABOLISM

All muscle and physical activity increases energy metabolism in the body compared with a state of rest. The increase in energy metabolism requires a greater uptake of oxygen and blood circulation. A relatively simple method of assessing energy metabolism is to measure *heart rate* (see Chapter 2, Section 2.10). There are commercially available and inexpensive personal monitoring instruments for recording heart rate which can store data for a complete working day or longer (Figure 6.9a–b). The advantage of this type of monitoring is that long-term monitoring can be carried out without disrupting the natural activity. The disadvantage is that the heart rate is also influenced by factors other than physical activity, for example, stress. This influence of mental processes may lead to substantial false estimates of the physical activity level at low levels, for example, in sedentary work. To acquire more reliable measurements of energy metabolism, one can measure oxygen consumption from the air inhaled (see Chapter 2, Sections 2.6 and 2.10, and Figure 6.9d). Sophisticated monitoring methods such as these are more suited to expert users.

6.12.6 MEASURING MUSCLE ACTIVITY

Muscle activity can be monitored with the aid of electrodes recording electrical activity from the action potentials, so-called *electromyography* (EMG). Today there is commercially accessible monitoring equipment for recording surface EMG (Figure 6.10). Portable equipment is also available that can store recorded activity for many hours. A simplified version of equipment of this kind usually forms part of so-called biofeedback equipment used to train individuals to relax their muscles, usually the trapezius.

Normally the total activity from the action potentials of many motor units is recorded with electrodes fixed to the skin outside the muscle, so-called surface EMG (Figure 6.11). This method is most suitable for surface muscles, for example, the trapezius or the flexors and extensors of the wrist on the forearm. The graphs that are shown are complex and should be compared with EMG registrations under special controlled conditions, for example, in a maximal voluntary contraction (MVC). EMG activity during work can then be compared with activity in the MVC measurement and expressed as a percentage of this (MVC%). Performing maximum muscle contractions may involve a certain risk

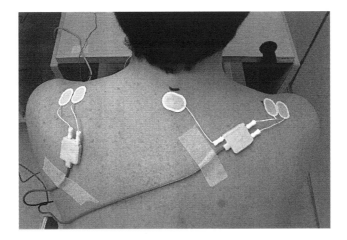

FIGURE 6.10 EMG recording of activity in the trapezius muscles. Photo: Allan Toomingas.

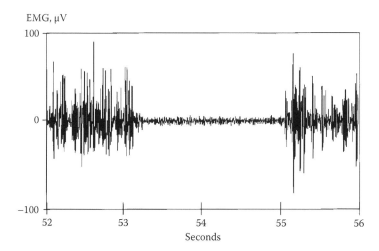

FIGURE 6.11   Surface EMG from the trapezius muscle including 2 s muscle rest (micropause). Photo: Göran Hägg.

of damage, which is why people often instead often use what is called a reference contraction, for example, holding a kilo weight with outstretched arm if the measurement applies to the trapezius muscle. The EMG activity during the work is then compared with the activity during the reference contraction (REF%). We are often interested in recording muscle rest, that is to say, that the EMG activity is zero, or near to zero.

The advantage of EMG registration is that it is a technically well-developed method of measuring muscle activity. Measurement records activity from a limited area (a few cm²), which is an advantage as it is possible in that way to know which muscle is being measured. One disadvantage, however, is that only a small part of the body's muscle activity is visible. Most movements and all work, however, demand coordinated activity from many different muscles. Another disadvantage is that with surface EMG it is only possible to measure activity in muscles near the surface. Moreover, the registrations may easily be subject to disturbance from, electrical devices, for example. In order to achieve high quality in the measurements, the registration should be quality checked, which can be time consuming if it is a question of a large number of measurements.

With the aid of thin needle electrodes pushed into the muscle, it is possible to record EMG activity from individual muscle cells. Activity patterns can be studied in individual motor units, for example, breaks with muscle rest. Such a zero activity can be very short (seconds), so-called EMG gaps. Measurements with needle EMG are very demanding and best suited to laboratory studies.

6.13   WHAT THE LAW SAYS ABOUT PROLONGED LOW-INTENSITY, SEDENTARY, AND STATIC WORK

Prolonged sedentary and low-intensity static work is not explicitly mentioned or covered by EU directives. Nor do the different national work environment laws take such work into consideration. There are, however, some EU Directives at a more general level that are applicable to such work. From the EU Council Directive 89/391/EEC the following general directives regarding the organization of work can be emphasized [EUR-Lex 1989] (italics mine).

1. "The employer shall take the measures necessary for the safety and health protection of workers, including prevention of occupational risks and provision of information and training, as well as provision of the *necessary organization* and means. The employer shall be alert to the need to adjust these measures to take *account of changing circumstances* and aim to improve existing situations" (Section II, Article 12).

2. "The employer shall implement the measures on the basis of the following general principles of prevention: A) adapting the work to the individual, especially as regards the design of work places, the choice of work equipment and the choice of *working and production methods...*; B) developing a coherent overall prevention policy which covers technology, *organization of work*, working conditions ...; C) *giving appropriate instructions to the workers*" (Section II, Article 6).

3. "The employer shall ensure that *each worker receives adequate safety and health training*, in particular in the form of information and instructions specific to his workstation or job" (Section II, Article 12).

4. "It shall be the responsibility of each worker to *take care as far as possible of his own safety and health ...*" (Section III, Article 13).

These articles stress the responsibility of the employer to organize the work and choose production methods that protect the health of the workers. Logically, this must also cover health risks from prolonged, sedentary, and static work. Employers should therefore organize work and design production methods bearing this in mind and informing and training their workers in how to avoid the risks. The workers have a responsibility to work in a way that is healthy.

Computer work has become very common in today's working life, as is described in Section 6.2. Computer work is, therefore, probably the most common setting for sedentary work. From the EU Council Directive 90/270/EEC regarding work at computers, the following relevant directives can be highlighted from Section II [EUR-Lex 1990] (italics mine).

1. "Employers shall be obliged to perform an analysis of workstations in order to evaluate the safety and health conditions to which they give rise for their workers, particularly as regards possible risks to eyesight, *physical problems* and problems of mental stress. Employers shall take *appropriate measures to remedy the risks* found ..., taking account of the additional and/or combined effects of the risks so found" (Article 3).

2. "The employer must plan the workers' activities in such a way that daily work on a display screen is *periodically interrupted by breaks or changes of activity* reducing the workload at the display screen" (Article 7).

3. "Workers shall receive information on *all aspects of safety and health* relating to their workstation, in particular information on such measures applicable to workstations as are implemented under Articles 3, 7 and 9.

Every worker shall also *receive training in use of the workstation* before commencing this type of work and whenever the organization of the workstation is substantially modified" (Article 6).

The risks of prolonged sedentary work were not generally recognized when this Directive was formulated in the 1980s. But the "physical problems" stressed in Article 3 may be applicable to risks from prolonged sedentary work. Organizing computer work with regular interruptions by way of breaks or other activities, as stated in Article 7, is one way of reducing such risks. It could also be argued that the Directive gives support to the suggestion that workstations should be equipped with height-adjustable desks, allowing the worker to alternate between seated and standing computer work in order to decrease sedentary work. Similar to Directive 89/361 above, information to computer workers about health risks and training in good working techniques is stressed here too.

The Directive also stresses aspects that influence the risk of prolonged static load on the musculoskeletal system, mainly in the neck and upper extremities. In the Annex, the Directive specifies requirements on the software, the environment and equipment at the workstation, including the furniture. These measures could reduce the risk of static load on the neck and upper extremities of the computer user. Adjustability and flexibility are stressed regarding the chair: "The chair shall be stable and allow the operator easy freedom of movement and a comfortable position" (Annex). The Directive also specifies the need for an eye test, and corrective glasses if necessary, before commencing computer work and at regular intervals thereafter (Article 9). These actions can also reduce the risk of static load on the neck and shoulders.

The International Standards Organization (ISO) has formulated a large number of quite detailed standards regarding different aspects of computer work stations [ISO 2010]. Standards have the state of "recommendations," not of law.

6.14 SUGGESTIONS FOR IMPROVEMENTS

6.14.1 ORGANIZATION OF WORK

Basically, work should be organized so that both physical and mental variation and recovery are built in. Uniform and monotonous loads should be avoided. Special care should be taken to ensure that such jobs are not done under time pressure or other stressful conditions. The combination with the high demands for precision should also be avoided, for example, as regards hand–eye coordination. If the work by its nature involves adverse loads of this kind, it should be alternated with other work tasks which are more favourable from these aspects and/or have breaks at regular intervals. The more adverse loads found in the work, the more important it is to limit the time workers are exposed to it, and it is even more important that the work is interrupted for variation and recovery. A ground rule is that it is better to have many short breaks than few long ones. In, for example, intensive computer work, breaks of 1–5 min several times an hour have proved to be advantageous also for productivity.

Both the knowledge base of physiology and studies from working life support the idea that loading that is "optimal" is best. "Optimal" applies then to both the physical and mental intensity of the load, its frequency and variation, and its duration (see Chapter 2, Section 2.12; Chapter 3, Section 3.14; Chapter 5, Section 5.8; Chapter 6, Section 6.11; Chapter 7, Sections 7.8 through 7.10, Chapters 9 and 10). Often the most effective means of achieving physical and mental variation and recovery is to alternate between different work tasks during the working day. Recovery does not need only to mean that the worker "rests" and is inactive. Carrying out other tasks or working in a different way may be recuperative for the muscles of the shoulders, for example. To do something more routine or more manual may be recuperative after work that has required great mental concentration. One could perhaps sort the post, make the afternoon coffee for the department, tidy one's desk, or tidy up around the workplace. One can alternate between routine tasks and meetings or personal in-service training or instructing new colleagues.

The work organization solutions for achieving variation are various. The degree of variation can be more or less extensive. One common method is the so-called *job rotation*. This implies that on a regular basis, usually several times a day, one alternates between different tasks of a similar nature. It might, for example, be a question of changing between stations on an assembly line where various parts are being fitted to a car. This may provide some variation in loads as regards, for example, work posture, force development in muscles, precision demands for hand – eye coordination, compared with what would be the case if one were to install the same part all day long. Another example is to alternate between sitting at a checkout in a supermarket where the customers' purchases come from the right or from the left respectively. Such variation is often rather limited, however. Often the musculature of the neck/shoulders is loaded in a similar way, as the different jobs between which the worker is alternating all require intensive work involving hands and arms.

A higher degree of variation is often desirable, and may be achieved using the so-called *job enlargement* or *job enrichment*. This is where alternation between tasks or new tasks of a completely different character is introduced. In the example with the assembly line, it might be a question of fetching parts from the stores, conducting quality control on the finished work, or planning next week's staffing of the line. In the example with the cashier in the supermarket, it may be a question of stacking shelves, price-marking goods, decorating and organizing signs or planning and ordering goods from the wholesaler. The examples describe the addition of tasks with an ever-increasing degree of mental complexity. More comprehensive variation often requires a broader competence than if a worker is carrying out one task. What the "optimal" variation is between the uniformly monotonous and the complex and demanding is different for different individuals and during different phases of their professional life. Further training and skills development are therefore important factors in making variation in work possible. Good organization of work provides scope and development opportunities for everyone to find the variation that is "optimal." There is no sharp dividing line between job rotation—job enlargement—job enrichment, but it is a question of differing degrees and characters of variation.

A different example is job exchange in which on a regular basis, usually with some days or weeks in between, the worker changes between different jobs, workplaces, and/or employers, for example, in an employers' circle. There are examples from health care, where nurses alternate between providing medical advice over the telephone and traditional nursing in a hospital. It may also be a matter of changing between completely different professions, for example, a (part-time) farmer who supplements his income with work as a salesperson at a local supermarket.

6.14.2 THE LEVEL OF ACTIVITY AT WORK AND IN CONNECTION WITH WORK

If the work is low intensity and sedentary by nature, all opportunities should be taken to add operations with favourable loading moments and an increase in energy metabolism. Arranging meetings on the move or having standing coffee breaks is a minor contribution. Sharing the responsibility for cleaning the workplace is another. The workplace is the place, apart from the family, to which most people regularly return and where they have a social network. Make use of the work group to motivate and stimulate good physical activity, for example, by distributing pedometers and establishing targets for the number of steps walked per week. The workplace may also join some kind of health-promoting activity through occupational health services or acquire fitness certification for their group. A physical fitness hour with a selection of activities during paid working hours is another well-justified solution.

The workplace can also facilitate physical activity in connection with commuting, for example, by organizing places to leave bicycles which are secure and protected from the weather.

6.14.3 WORK EQUIPMENT

Work equipment, for example, desks and chairs, should be designed so that they *allow for physical variation* and *recovery.* An example of this is a desk, for example, for computer work, which can easily be raised or lowered for variation between seated and standing work.

It is vital that the working environment in general does not cause further adverse loads, for example, glare from windows or light fittings, or stressful noise from a ventilation unit. Particular problems can arise in so-called open office landscapes where a large number of potentially stressful elements may exist.

Employees should know how equipment is used and be aware of suitable working technique. It is also necessary for them to understand why it is important and what health risks there are at work. One should be aware that such knowledge does not last, and needs to be refreshed from time to time.

There is essential, concrete ergonomic advice about workplaces and choice of equipment in connection with computer work published in literature and on various web sites [Toomingas 2007].

6.14.4 THE INDIVIDUAL

The keywords for the individual are to use a working technique that *makes use of the opportunities* for *variation* and *recovery.* Variation and recovery apply to both physical and mental work. In order to avoid work involving prolonged static load, workers should vary their work posture and movements frequently. It is also important to *relieve the load* on the arms and hands and the weight of any tools and materials. In computer work, it is a good idea if the forearms can rest on armrests or the desk top when working at a keyboard and with the mouse. One should also avoid prolonged work postures that deviate from the neutral position of the joints, for example, looking up for a long time or turning one's neck to the side. It is therefore important to adjust equipment in the correct manner so that adverse loading on the body is minimized.

Several concrete pieces of advice about working technique in connection with computer work can be found on various web sites [Toomingas 2007].

One way of achieving physical variation in computer work is to alternate between sitting and standing. Computer workers who stand for 4 h a day instead of the "normal" 2 h increase their energy metabolism by the equivalent of a weight loss of ~2.5 kg/year. As the work is of such low intensity that the body does not get the necessary stimulation for reconstruction, all other opportunities for increasing energy metabolism and dynamic load should be utilized. One might, for example, walk up the stairs instead of taking the lift or escalator. Cleaning or carrying out other practical tasks at the workplace not only provides a necessary addition to the load, it also increases variation both physically and mentally.

In most low-intensity and sedentary work, however, there is a practical limit to how much the work tasks can be varied or the energy metabolism increased. More physically demanding and energy-intensive activities outside work must then supplement this. Taking a lunchtime walk, maybe together with a group of friends, is one example. Commuting is another important opportunity that should be used, as they occur frequently. Walking or cycling instead of taking the car is also good for the environment.

A common recommendation is that every adult should be physically active for a total of at least 30 min everyday (possibly divided into 10 min sessions) at a level that corresponds to a brisk walk (~50% of maximal aerobic capacity) [SNIPH 2010]. Lunchtime walks and walking to and from work fit in here very well. What is more, two or three times a week for about 45–60 min one should exert oneself so as to get really out of breath and sweaty (~75% of maximal aerobic capacity). Activities which are pleasurable and which suit personal tastes and interests should be chosen. The choice may be between traditional gymnastics, aerobics, jogging, dance, hard work in the garden, and the like.

6.14.5 SOCIETY

Society should, through its various organizations, authorities, and functions, strive to provide a working life that promotes physical and mental variation and recovery, together with optimum physical load and energy metabolism. An example of this is training and information in such issues of public health for both individuals and also, for example, social decision-makers, CEOs, and representatives of business. Schools should take up these questions early in childhood and emphasize in their timetabling the importance of regular physical activity and good working technique as well as the avoidance of sitting still for prolonged periods. Through different financial incentives it is possible to stimulate individuals and companies to promote physical activity. Society's various organizations should also stimulate ways of promoting physically active leisure time.

The local community can stimulate and facilitate active commuting by, for example, designing footpaths and cycle tracks that are safe from traffic, particularly on roads that connect to major workplaces.

6.15 SUMMARY

Sedentary, prolonged low-intensity work is common, for example, in computer jobs. Despite their low load, jobs of this kind may cause disorders, particularly in the neck and upper extremities. Comprehensive documentation exists providing various hypotheses about causal mechanisms behind the origins of such disorders. Impairments may occur in blood flow to the muscles, tendons, and nerves. An uneven distribution of load between different parts of the muscle may lead to overload in particularly vulnerable parts. Working with one's hands also activates muscles in the neck and shoulders, which can then be subjected to long-term load, particularly if muscle tensions induced by stress appear. Relaxing breaks and a variation in work posture and movements, or, even better, variation involving other activities entirely, may reduce the risk of problems. Too low of a load over a prolonged period may lead to damaging metabolic processes and a lack of buildup of capacity in the tissues and organs of the body, which in the long term may lead to various more or less serious health conditions, for example, in the cardiovascular system. Prolonged, sedentary low-intensity work should be alternated with other more mobile activities and be supplemented by regular physical exercise.

REFERENCES

Ainsworth, BE. et al. 2000. Compendium of physical activities: An update of activity codes and MET intensities. *Med Sci Sports Exerc* 32(Suppl 9):498–504.

Åstrand, PO., Rodahl, K., Dahl, H., Strömme, S. 2003. *Textbook of Work Physiology. Physiological Bases of Exercise.* 4th ed. Champaign: Human Kinetics.

Bernard, B. 1997. *Musculoskeletal Disorders and Workplace Factors: A Critical Review of Epidemiological Evidence for Work-Related Musculoskeletal Disorders of the Neck, Upper Extremity and Low Back.* Washington: CDC-NIOSH.

Borg, G. 1998. *Borg's Perceived Exertion and Pain Scales.* Champaign: Human Kinetics.

Côte, P. et al. 2008. The burden and determinants of neck pain in workers. *Spine* 33:S60– S 74.

Dept. of Health. 2004. *At Least Five a Week: Evidence on the Impact of Physical Activity and its Relationship to Health. A Report from the Chief Medical Officer.* London: Department of Health.

Ekblom-Bak, E., Hellenius, ML., Ekblom, B. 2010. Are we facing a new paradigm of inactivity physiology? *Br J Sports Med* 44:834–835.

EUR-Lex. 1989. *Council Directive of 12 June 1989 on the Introduction of Measures to Encourage Improvements in the Safety and Health of Workers at Work* (89/391/EEC). http://eur-lex.europa.eu/en/index.htm

EUR-Lex. 1990. *Council Directive of 29 May 1990 on the Minimum Safety and Health Requirements for Work with Display Screen Equipment* (90/270/EEC). http://eur-lex.europa.eu/en/index.htm

Eurofound. 2010. *Fifth European Working Conditions Survey—2010*. European foundation for the improvement of the living and working conditions. http://www.eurofound.europa.eu/surveys/ewcs/2010/index.htm

European Agency. 2005. *Expert Forecast on Emerging Physical Risks Related to Occupational Safety and Health*. Luxembourg: European Agency for Safety and Health at Work.

Healy, G. et al. 2007. Objectively measured light-intensity physical activity is independently associated with 2-h plasma glucose. *Diabetes Care* 30:1384–1389.

Huang, GD., Feuerstein, M., Sauter, SL. 2002. Occupational stress and work-related upper extremity disorders: Concepts and models. *Am J Ind Med* 41:298–314.

Hägg, G. 2000. Human muscle fibre abnormalities related to occupational load. *Eur J Appl Physiol* 83:159–165.

ISO. 2010. *Ergonomic Requirements for Office Work with Visual Display Terminals (VDTs) ISO 9241*. International Organization for Standardization. http://www.iso.org/iso/home.html

Johansson, H., Windhorst, U., Djupsjöbacka, M., Passatore, M. 2003. *Chronic Work-Related Myalgia. Neuromuscular Mechanisms Behind Work-Related Chronic Muscle Pain Syndromes*. Gävle: Gävle University Press.

Järvholm, U., Styf, J., Suurkula, M., Herberts, P. 1988. Intramuscular pressure and blood flow in supraspinatus. *Eur J Appl Physiol* 58:219–224.

Katzmarzyk, P., Church, T., Craig, C., Bouchard, C. 2009. Sitting time and mortality from all causes, cardiovascular disease, and cancer. *Med Sci Sports Exerc* 41:998–1005.

Lynch, L. 2010. Sedentary behavior and cancer: A systematic review of the literature and proposed biological mechanisms. *Cancer Epidemiol Biomarkers Prev* 19:2691–2709.

Matthews, C. et al. 2008. Amount of time spent in sedentary behaviours in the United States 2003–2004. *Am J Epidemiol* 167:875–881.

Moradi, T., G. Gridley, Björk, J., Dosemeci, M., Berkel, HL., Lemeshow, S. 2008. Occupational physical activity and risk factor for cancer of the colon and rectum in Sweden among men and women by anatomic subsite. *Eur J Cancer Prev* 17:201–208

Morris, JN., Heady, JA., Raffle, PAB., Roberts, CG., Parks, JW. 1953. Coronary heart disease and physical activity of work. *The Lancet* 1053–1057 + 1111–1120.

Mummery, K., Schofield, G., Steele, R., Eakin, E., Brown, W. 2005. Occupational sitting time and overweight and obesity in Australian workers. *Am J Prev Med* 29:91–97.

Palmer, K. and Smedley, J. 2007. Work relatedness of chronic pain with physical findings—A systematic review. *Scand J Work Environ Health* 33:165–191.

Savinainen, M., Nygard, CH., Ilmarinen, J. 2004. A 16-year follow-up study of physical capacity in relation to perceived workload among ageing employees. *Ergonomics* 15:1087–1102.

Sluiter, J., Rest, K., Frings-Dresden, M. 2001. Criteria document for evaluating the work-relatedness of upper-extremity musculoskeletal disorders. *Scand J Work Environ Health* 27 (Suppl 1).

SWEA. 2010. *The Work Environment 2009*. Stockholm: Swedish Work Environment Authority, pp. 58–69. http://www.av.se/dokument/statistik/officiell_stat/ARBMIL2009.pdf

SNIPH. 2010. Physical activity in the prevention and treatment of disease. The Swedish National Institute of Public Health. Report R 2010:14.

Toomingas, A. 2007. *Computer Work*. Stockholm: Swedish Work Environment Authority. http://www.av.se/dokument/inenglish/themes/computer_work.pdf

van Uffelen, J. et al. 2010. Occupational sitting and health risks. *Am J Prev Med* 39:379–388.

Veiersted, B., Weestgaard, RH., Andersen, P. 1993. Electromyographic evaluation of muscular work pattern as a predictor of trapezius myalgia. *Scand J Work Environ Health* 19:284–290.

Visser, B. and van Dieën, JH. 2006. Pathophysiology of upper extremity muscle disorders. *J Electromyogr Kinesiol* 16:1–16.

FURTHER READING

Arnetz, B. and Ekman, R. 2006. *Stress in Health and Disease*. Weinheim: Wiley-WCH.

Delleman, N., Chaffin, D., Haslegrave, C. 2004. *Working Postures and Movements—Tools for Evaluation and Engineering.* Boca Raton, FL: Taylor & Francis.

Kuorinka, I. and Forcier, L. (eds). 1995. *Work-Related Musculoskeletal Disorders (WMSDs): A Reference Book for Prevention.* London: Taylor & Francis.

Wilmore, JH., Costill, DL., Kenney, WL. 2008. *Physiology of Sport and Exercise*. Champaign: Human Kinetics.

7 Work with High Levels of Mental Strain

Bo Melin

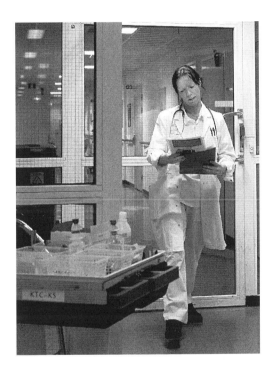

Photo: Joakim Romson

CONTENTS

Victoria is a doctor, aged 30. Her work in a care profession means that she must be responsive and take a large responsibility for patients' lives and safety. She often has to make quick and rational decisions both on her own initiative and together with other people. The demands from patients and the fear of making a mistake often result in a feeling of concern that will not go away even after the end of a day's work. Her experience of stress, felt in both body and mind, is very tangible.

Working with people, and not least working with people in distress, often involves considerable mental strain. Staff in not only the care professions, but also in schools, and those working at call and contact centres, in a reception, in shops, and restaurants, are occupational groups burdened with considerable social and mental demands. Within these occupations, staff have to devote energy to, for example, restraining spontaneous feelings and maintaining detachment in their relationship with patients, students, clients, or customers.

Being subject to heavy psychological demands in our working environment does not in itself need to imply anything negative for our own health, but may on the contrary provide a stimulus, provided the stresses feel manageable and are of relatively short duration, and provided there is an opportunity for recuperation. But when the demands become too great, or when we experience prolonged frustration for other reasons, a number of negative mental and physical reactions may occur.

Sometimes Victoria has to operate. In surgery, experience, precision and "high technology" are often characteristically used in conjunction. Even when she is operating, the mental strain is pronounced; but it is of a rather different character, involving less emotion than when out in the ward. This is where the strain has more to do with concentration, a mental or cognitive involvement which must not be disturbed. If, nevertheless, a disturbance does occur, there is an increased risk of making a mistake, which would be very serious in the case of an operation. Many occupations involve great demands on precision and concentration—for example, the work of welders, precision tool makers, dentists, and slaughterhouse workers. A high level of cognitive involvement that cannot be disturbed is required, for example, of interpreters, call centre operators, and actors.

7.1   FOCUS AND DELIMITATION

This chapter has a biopsychosocial framework; it describes psychological exposures and reactions to them. The chapter differs from other chapters insofar as here the focus is on exposures that only individuals themselves can describe and express. Here there are no objective exposure limits for the relevant exposures. Mental exposures are in a sense invisible, and are made visible and recordable with the aid of various methods such as scales, questionnaires, brain scanning techniques, psychophysiological techniques, and different types of experiments. It is the reactions to the exposures that are recordable and can be measured both subjectively and physiologically. By way of introduction, we describe not only what a stressor is and the differences between physical and mental stressors, but also their similarities in the stress reaction itself. *Allostatic load* is important in this context. Thereafter, we give an account of the biology of the stress reaction to subsequently deal with mental strain in two partially different senses. On the one hand, we deal with the mental strain resulting from how the psychosocial work environment is configured in general, that is, how work is organized, how time pressures and feelings in interpersonal relationships impact on the person as an organism (e.g., Victoria as a doctor in the accident and emergency department and out in the ward). On the other, we deal with the mental strain caused by more specific cognitive demands in the work. In other words, how people and their brains interpret and process specific information and how they react to and manage this information (e.g., Victoria during surgery). In our technological age, the configuration of the working environment is largely a question of how we use and function together with technology. Finally, the relationship between mental load and muscle activity is illustrated, as well as what the Work Environment Act has to say about mental load.

In this chapter, the following questions will be addressed:

- What is the difference between physical and mental load in working life?
- How does mental load and stress in working life originate?
- What happens in the body under mental strain and stress?
- Can mental load and stress at work be harmful?
- Can we measure the degree of mental strain and stress?
- Has mental load in working life changed over the years?
- What does the Work Environment Act have to say about mental load in working life?

7.2   INCIDENCE OF WORK WITH HIGH MENTAL LOAD

The mental load on doctors is often described as high, and in a population of Swedish doctors almost three quarters stated that the work is mentally demanding [Ohlin 2001]. The mental load derives partly from the work itself with patients, but also in large part from comprehensive organizational changes and new IT-related administrative operating systems that have to be run by the doctors themselves. The doctor's situation in this way has similarities to the working situation of others who work with people in vulnerable situations, and who are experiencing similar organizational changes. In the Western world it is work within the nursing, school, and care sectors that provides the most jobs. In 2006, it was estimated that 68.6% of those employed in Europe were working in these sectors, and only 25% in industry [Eurostat 2008]. The trend in Europe is to a continued percentage rise within nursing, schools, and care, and a continued percentage decrease in industry. In Sweden in the 1990s, and at the beginning of the millennium, it was within these sectors that absence due to illness, as well as the length of absence, was most pronounced, and where mental symptoms and injuries due to strain comprised a very high proportion of long-term absence due to illness. The European Union has identified that "work-related stress is, after back pain, Europe's biggest work-related health problem." It is estimated that these work-related health problems cause more than half of all sick leave in Europe [European Agency for Safety and Health at Work 2004].

7.3   STRESSORS AND STRESS REACTIONS

Time pressures and high psychological demands are examples of *stressors*, or conditions that exist in many occupations (read more about the psychosocial working environment later in this chapter). If Victoria experiences these factors emotionally or cognitively as time pressures or as demands that are too great, they constitute stressors. These are usually called *mental stressors*. Stressful heat, cold, and noise are also common stressors, but these are usually assigned to the category *physical stressors*.

Victoria lives in a changing work environment, with both time pressures and considerable demands, which in their turn make great demands on the *adaptability* of her body or organism and imply that in every "new" situation she has to react in a way that is functional for her. The *stress reaction* is an example of our ability to adapt. The stress reaction is produced in stressful situations and these are a normal physiological reaction resulting from a stressor.

A *mental or cognitive stressor* is as a rule qualitative in kind (stress that has to be interpreted by the individual) so that it puts strain on the memory or demands continuous attention, rapid reactions, or contact with other people. If the work is closely controlled and inflexible, or involves too much responsibility or competition, it may lead to high mental strain. The situation can be exacerbated if unclear or contradictory expectations are imposed on the employee. The same applies to too great a quantity of work and a lack of access to information, as well as changes occurring in the work itself.

7.3.1   Are There Any Differences between Mental and Physical Stressors?

As regards exposure, it may seem self-evident to distinguish between physical and mental stressors. Stress reactions caused by physical stressors such as noise, vibration, heat, and cold can disturb the internal balance of the body so that a stress reaction results. Physical stressors disturb the internal balance of the body, regardless of how we experience them mentally. It can be said that physical stressors have an impact on the internal environment of the body without higher cognitive functions needing to be employed (read more about cognitive functions later in this chapter). As regards the stress reaction itself, it is interesting that the division into physical and mental stress reactions is not as obvious. The allostatic model (see Fact Box 7.1) describes a harmful course of events on the basis of the concepts of allostasis and allostatic load (Figure 7.1) [McEwen 2000a,b]. *Allostasis* means the process that ensures that the body's physiological systems are in balance, and that homeostasis is thereby maintained during varying external conditions. These external conditions may be of different kinds—physical as well as mental/psychosocial. Allostasis has a primarily protective function for the body, but allostatic processes can also act to the

---

FACT BOX 7.1

During the 1990s, Bruce McEwen, Professor of Neuroendocrinology at Rockefeller University, introduced the concepts of "allostasis" and "allostatic load" in relation to homeostasis. In the scientific literature, allostatic load has, in part, replaced the concept of stress, and allostasis is seen as a complement to homeostasis. The concept of allostasis is drawn more widely than homeostasis and emphasizes, among other things, the importance of understanding the mechanisms for recuperation.

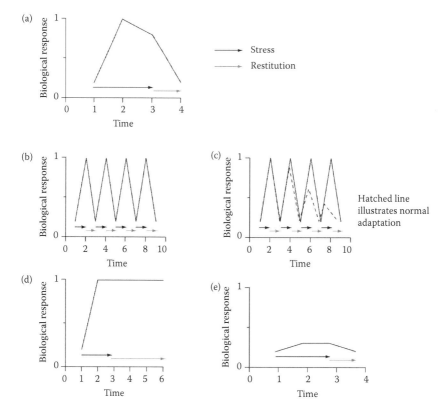

FIGURE 7.1 According to McEwen's model, a normal or sound reaction is to be found, and four types of situations which can lead to allostatic load, resulting in less sound reactions over time Modified from McEwen [2000a] by Melin [2003]. (a) Normal stress reaction, with mobilization and easing off/recuperation when the stressor has ceased. (b) Repeated reactions as a result of challenges precipitated by repeated stressors. (c) Failure to adapt to repeated challenges. Adaptation here (the circled line) can save the body from excessive stress hormones or load, for example. One example of this is appearing in public, which initially can trigger anxiety and stress symptoms, but when repeated can become a habit which does not feel at all disagreeable or burdensome. In the case of phobias, for example, a phobia of spiders, each time when a spider appears a strong physiological and mental reaction is triggered. In phobia treatment tools are provided to manage the stressor (the spider), whereby more adequate mental and physical reactions occur on sighting a spider. A certain habituation can also occur as a result of exposure to physical stressors (e.g., noise), but it is seldom suitable to accustom oneself or adapt oneself to. The physical stressors that have a considerable impact on homeostasis should preferably be avoided completely. (d) Inability to "shut off" allostatic response. This is the largest category of allostatic load as regards the existing examples. Not everyone's blood pressure recovers after acute stress. Failure to reduce HPA activity (read more about the HPA axis later in this chapter) in the evening is a kind of allostatic load. A perhaps unexpected consequence is that women may have a reduced bone-mineral density as a result of depression. This may be due to allostatic load with raised hormone levels, which is linked to depression, also causing chronically reduced calcium levels. (e) Inadequate allostatic response may occur, which triggers a compensatory increase in other allostatic systems. If, for example, adrenalin is not secreted as a response to stress, then the sensitivity to autoimmune and inflammatory disturbances will increase (Fact Boxes 7.2 and 7.3).

detriment of the body and produce harmful effects if they do not disappear within a reasonable period. In other words, they give rise to *allostatic load.*

The normal allostatic reaction, for example, the stress reaction, is triggered by a stimulus and is subsequently maintained over an appropriate time interval so as finally to cease (illustrated in Figure 7.1a). Four situations (illustrated) are thought to promote negative allostatic load (Figure 7.1b–e); (b) repeated and frequent stress; (c) inability to adapt to stress and adequately reduce the stress response; (d) inability to shut off the stress response after a stressor has ceased; (e) in an interrupted stress response when, for example, the stress hormone level is not raised in a stress situation and instead other compensatory physiological responses ensure that allostasis is maintained [Melin 2003, pp. 237–238].

The prolonged load to which these situations give rise is exhausting for the body. The result of maintaining allostasis under such circumstances is increased vulnerability in those organs and systems which act to restore homeostasis, and as a result physical and mental symptoms may arise.

According to the reasoning about allostatic load that has been given, there is no difference whether, in her work as a doctor, Victoria is subjected to mental or physical stressors. From an allostatic perspective, there is therefore no difference between a physical and mental stressor. However, [McEwen 2000b] considers that a stress reaction is absolutely necessary in situations requiring physical activity, but that it is scarcely functional when we are subjected to psychosocial stress, with no element of physical activity. Many jobs today contain few or no elements of physical activity.

7.3.2 ABSENCE OF RECUPERATION

McEwen's model, like previous stress models, emphasizes that energy mobilization is something healthy, sound, and normal in situations requiring energy. At the same time, the model focuses on the fact that problems only arise when energy is mobilized constantly without any periods of recovery. Perhaps the individual has been giving all they have for months and years without sufficient recuperation. In prolonged stress and allostatic load, the balance has shifted between the body's *catabolic* ("breaking down") and *anabolic* ("building up") conditions towards the catabolic end, which has an impact on the constantly ongoing remodeling process described in Chapter 6, Section 6.13. The result may be weakened structures and impaired healing processes. The situation can be exacerbated by a lack of recuperation, particularly in cases of prolonged sleep disturbance (Fact Boxes 7.2 and 7.3).

7.4   STRESS REACTION: A PHYSICAL RESPONSE TO MENTAL EXPOSURES

A stressor, mental or physical, gives rise to a stimulus causing the body's homeostasis—its constant internal environment—to be disturbed. The stress reaction is the adaptation that is made, *neurally* and *endocrinally*, in order to restore the homeostatic condition in the body when a stressor has thrown it out of balance.

FACT BOX 7.2

In prolonged energy mobilization, the so-called hypothalamic–pituitary–adrenal (HPA) system facilitates the body's preparedness to cope with situations that demand energy. This system is activated by stress, and initially provides an increase of cortisol in the blood (the body's own cortisone) and enhances the capacity of the immune defense system. If the system grows tired, no mobilization/activation takes place. The cortisol level in the blood is then low and does not increase (is not mobilized) to match external stress. In normal cases, the cortisol level in the blood is high in the morning—but with fatigue it is low in the morning instead—and remains unchanged throughout the day instead of dropping towards the evening [Kristensen et al. 2011].

FACT BOX 7.3

In animal studies [De Kloet et al. 1998], it has been observed that raised cortisol levels lead to impaired cell function or cell death, respectively, in certain areas of the hippocampus (e.g., a deep-seated structure in the limbic system that is important for our short-term memory). Corresponding finds have been made using a magnetic resonance imaging (MRI) scanner on people with post-traumatic stress disorder (PTSD) and in adults who have suffered known serious abuse as children. This in its turn correlates with impaired verbal memory found in neuropsychological tests. The two types of receptors for corticosteroids are: mineralocorticoid receptors (MR) and glucocorticoid receptors (GR). MR are to be found to a great extent in the hippocampus and in other areas of the limbic system. They are activated by basal HPA activity, which follows a diurnal rhythm and has a bonding capacity, that is, 6–10 times stronger in relation to corticosteroids than GR. GR are in their turn also localized in different areas of the brain, primarily in the hippocampus, the hypothalamus, the amygdala, and the prefrontal cortex [Nordling 2003, pp. 18–19].

When HPA activation increases, for example, in cases of acute stress, GR are involved to a greater extent, and then a suppression of the HPA axis usually follows [McEwen 2006]. If the system is subjected to prolonged stress, however, the effect is quite different. The hippocampus is the structure that has primarily been studied, and seems to be affected by prolonged high levels of stress hormone. This should be seen against the basis of the large proportion of both MR and GR receptors (described above) that can be found within the structure. In chronically raised levels of cortisol, signal transmission in the hippocampus becomes damaged, and outgoing communications are reduced. The consequence is that the inhibition of corticotropin-releasing hormone (CRH)-producing neurons in the hypothalamus is disturbed, and in this way the release of cortisol is not reduced. One hypothesis is that this may be a possible explanation for the cognitive disturbance observed in many patients with

poor concentration, memory, and over-sensitivity to many impressions. Even if
the patients themselves feel better, this cognitive disturbance seems to last for
a long period, causing problems and a long drawn-out rehabilitation [Nordling
2003]. You can read more about stress and the involvement of the brain in Fact
Box 7.4.

7.4.1 NEURAL ADAPTATION

On the *neural* level, the stress reaction leads to a greater activation of the sympathetic nervous system, and an inhibition, generally speaking, of the parasympathetic
nervous system. The effects of an increased sympathetic activation is that the heart
rate (HR), blood pressure, and blood sugar levels are raised, the pupils dilate, sweat
is secreted, and the external blood vessels contract. Researchers usually describe this
response or reaction as a necessary adaptation for action, either through attack (fight)
or flight. The reaction is intended to defend the body against an acute danger.

7.4.2 ENDOCRINE ADAPTATION

Endocrine adaptation occurs through an increased internal secretion of certain hormones, for example, adrenalin, norepinephrine, glucocorticoid, glucagon, prolactin,
and vasopressin. At the same time, the internal secretion of other hormones decreases,
for example, testosterone and, in the long term, also growth hormones.

   The neural and endocrinal changes together result in the body adapting to the
acute stress situation through increasing the availability of stored energy increases
and intensifying cardiovascular activity. The adaptation also means that the body's
development—that is, *anabolic functions* such as growth, reproduction, and
digestion—decline at the same time as the immune defenses are affected and cognitive acuity is raised.

7.5   STRESS REACTION IN TWO DIFFERENT SYSTEMS: THE SYMPATHETIC ADRENOMEDULLARY SYSTEM AND THE HPA AXIS

It is important to be aware that the dominating effects of the stress reaction result
from activity in two separate but not independent systems: the sympathetic adre-
nomedullary (SAM) system and the HPA axis (in somewhat older literature the
abbreviation PAC may be found). The central substances acting in the systems are
the classic stress hormones *adrenaline* and *norepinephrine*, which are collectively
called catecholamines (SAM) or *glucocorticoids*, primarily in the form of *cortisol*
(HPA). The best-known health problems associated with these systems affect the
cardiovascular system, such as, for instance, high blood pressure, heart attacks, and
cerebral infarctions. But prolonged activation or strain on these systems is also
assumed to be of significance for the development of diabetes (reduced insulin
sensitivity), infections (disturbed immune function), and cognitive impairment

(hippocampus degeneration, see Fact Box 7.4). The hormones adrenalin and norepi-nephrine dominate in situations requiring active exertion. High or markedly low cortisol levels (HPA) seem, on the other hand, to be related to a greater extent to situations when the individual is passive and helpless and where they are subjected to stress for a long period. Both systems are described in greater detail below [McEwen 2006].

7.6 AUTONOMOUS NERVOUS SYSTEM AND SAM SYSTEM

7.6.1 SYMPATHETIC AND PARASYMPATHETIC NERVOUS SYSTEM

The *autonomous* nervous system spontaneously controls activity in organs such as the heart, the adrenal medulla, the glands and the stomachic-intestinal canal, and activity in the smooth muscles, for example, around the blood vessels. The overriding function of the autonomous nervous system is to maintain a constant internal environment—homeostasis—and this is traditionally divided into the *sympathetic* and *parasympathetic* nervous systems. These are usually activated antagonistically. In the case of the heart and the intestines, for example, one system acts as a stimulus and the other as a retardant.

The activity in the *parasympathetic nervous system* helps, among other things, to reduce HR and increase activity in the stomach and intestines at the same time as producing contractions in the pupils as well as in the bladder and intestine. In the *sympathetic nervous system* there is a certain constant activity to maintain blood pressure and heart activity. Increased activity occurs in the system when the body is subject to stress and the internal homeostasis is disturbed. This may happen, for example, with considerable mental and emotional strain, but also with hypothermia, oxygen deficiency, muscular strain, or severe pain. The effects of increased sympathetic activation are raised blood pressure, HR, and blood sugar levels; dilated pupils; sweating; and contraction of surface blood vessels. Researchers usually describe this reaction as a necessary adaptation for action, either through attack (fight) or flight, and its purpose is to protect the body against urgent danger.

7.7 HYPOTHALAMUS–PITUITARY–ADRENAL AXIS

As a reaction to a mental stressor, the body also reacts—apart from the activity in the autonomous nervous system described above—with an increased production of corticosteroids secreted from the adrenal cortex. Corticosteroids are divided into mineralcorticoids which regulate the salt balance of the body, and glucocorticoids which, among other things, increase the amount of glucose (sugar) in the blood.

Glucocorticoids—the stress hormone cortisol is one—are internally secreted from the adrenal cortex as a result of physical or mental stressors. The internal secretion is preceded by activity in the HPA axis and its purpose is to maintain a physiological reaction primarily associated with acute stress.

Apart from the fact that the glucose level is raised, greater glucocorticoid secretion helps mobilize energy through activating a breakdown of carbohydrates, proteins, and lipids (blood fats). When the glucocorticoid level is raised, cardiovascular activity also increases at the same time as the anabolic processes associated with reproduction, growth, and immune defense slow down. Glucocorticoids also act as an anti-inflammatory. The glucocorticoid cortisol enters the blood stream within a few minutes after a stress stimulus, and bonds primarily (95%) to protein; the remainder circulates freely and bonds with different structures in the brain. In recent years, cortisol has been the subject of intensive study in relation to the brain. Important areas of the brain that have receptors for cortisol are the hippocampus, the amygdala, the pituitary gland, the hypothalamus, and the prefrontal cortex. The hippocampus is the structure that has attracted the greatest interest in research into the effects of stress on our brain structures. This is because it has a large number of receptors for cortisol, and is also involved in important cognitive processes (see Fact Box 7.4).

FACT BOX 7.4

STRESS AND THE INVOLVEMENT OF THE BRAIN

Several brain structures are involved in the HPA axis, and the internal secretion of glucocorticoids from the adrenal cortex is thus controlled by the brain. Information comes into the brain about homeostasis, and signals are integrated into the brain structure called the *hypothalamus*. Various structures of the brain which project to the hypothalamus include the amygdala and hippocampus in the limbic system within the cerebrum. Under stress, these areas receive information about somatosensory and *visceral* change, chemical levels in the blood and, moreover, information of a more psychological nature, such as the presence of threat, aggression, other emotions, and motivation. The base level of glucocorticoids that the body always needs is governed partly by the area in the hypothalamus called the suprachiasmatic nucleus. Information comes in here about *circadian* rhythms in the form of light changes during the day (see Chapter 8 on Diurnal Rhythm). The concentration of glucocorticoids follows this rhythmicity, and changes during the day so that the level is lowest when falling asleep and highest directly after waking up. It is therefore very important to monitor this rhythmicity in connection with measuring the hormone cortisol [McEwen 2006].

DO CHANGES IN THE HIPPOCAMPUS RESULT FROM PROLONGED STRESS?

Interesting pieces on structural changes in the hippocampus are reviewed [Alderson and Novack 2002], as well as [Sapolsky 2003; McEwen 2000b]. The research situation is such that no certain correlation has been proved in humans between high cortisol levels as a result of stress and neuron death (atrophy) in the hippocampus. There is some support from, among other sources, an MRI study of traumatized war veterans who were found to have smaller hippocampus volumes than corresponding healthy individuals [Gurvits et al. 1996; Gilbertson et al. 2002]. In animal studies, atrophy—that is, the withering away of the neuron—has been discovered in the hippocampus, which is an example of the harmful effect of prolonged, repeated, and chronically high levels of glucocorticoids. Studies also show that raised concentrations of cortisol can lead to harmful effects in the hippocampus in the form of increased vulnerability on the part of the neurons, a decline in neurogenesis—that is, a regeneration of neurons—and a smaller contact surface with other neurons through the filaments of the neurons, the dendrites, being affected [Nordling 2003].

7.7.1 THE QUESTION IS A VERY COMPLEX ONE

Researchers always emphasize the complexity in the process contributing to structural changes in the hippocampus (and also the prefrontal cortex) as a result of stress and the fact that, for example, cortisol is only one among many other processes affecting the harmful course of events. Moreover, there are divided opinions on why structural changes would arise in the hippocampus as a result of overexposure to cortisol.

Even low levels of cortisol have been shown to disturb the HPA axis and to produce negative effects on the hippocampus. This has been proved in certain cases of PTSD. The mechanism has, however, not been explained. Recently, research has drawn attention to the relationship between PTSD and the structural changes in the hippocampus associated with the symptoms. A twin study indicates that a smaller hippocampus volume may be the cause of greater sensitivity to PTSD. In this way, the size of the hippocampus may have a significant role even in the development of mental trauma and may thus be linked to an individual's vulnerability to stress. This can be compared with other theories that the structure undergoes morphological changes *after* a trauma and, therefore as a consequence of stress acts as a kind of "trauma memory." A great deal of research remains to be done before we can gain clarity into how the brain is affected by stress, and as the heading indicates, the question is a very complex one.

7.7.2 INTERPLAY BETWEEN THE SAM SYSTEM AND HPA AXIS, AND SALIVA CORTISOL RELATIONS TO HEALTH

How the interplay between the two systems occurs is not entirely understood, but it is probable that it is not a question of two independent systems. Research has shown, for example, that norepinephrine from the SAM system can also stimulate activity in the HPA axis. What is more, it has come to light that the same hormones regulating activity in the HPA axis (CRH) also affect the release of adrenaline and norepinephrine from the SAM system. The time it takes for cortisol to be secreted in a stress reaction is several minutes. This can be compared with the rapid catecholamines adrenalin and norepinephrine, which are released within seconds of exposure to a stressor. As a result of, among other things, this delay in releasing cortisol, several researchers have asked themselves what function cortisol really performs. Is the hormone part of the stress reaction in itself, or does cortisol have a longer term, protective effect on those processes arising in a stress reaction? Both explanations are usually put forward as probable, and possibly cortisol fulfills both functions (see Fact Box 7.5).

FACT BOX 7.5

(A) Cortisol mobilizes energy and inhibits anabolic processes that are *not immediately needed* (in acute stress) to handle the stressor, which may be regarded as a strategic adaptation at the precise moment of stress. (B) The immune defense is inhibited so as to protect the individual against the harmful effects of the stress reaction in a longer-term perspective (with protracted stress) [Nordling 2003, pp. 16–18].

In work-related studies, saliva cortisol is an often-used measure. Recently, [Kristenson et al. 2011] have reviewed this literature. They conclude that it is apparent that single measures of absolute concentrations of salivary cortisol, for most health-related variables, seldom give significant findings; deviation measures, in terms of diurnal deviations and/or laboratory stress tests seem to be more strongly and consistently associated with a number of factors, such as socioeconomic status, psychological characteristics, biological variables in terms of overweight and abdominal fat accumulation, and mental and somatic disease. Across disorders, the pattern related to ill-health/stress is generally characterized by a flatter diurnal cortisol curve, which in most cases is due to attenuated morning and/or increased evening levels, or a reduced response to a laboratory stress test. For some specific questions, single mean values seem to provide valuable information, but in all cases a careful design in terms of power and standardization is important. However, the authors conclude: "thus, salivary cortisol can be a useful biomarker in many settings, if caution is taken in the choice of methods used."

7.7.3 CONNECTION BETWEEN STRESS AND COGNITION

Stress research has long attempted to study the connections between various psychosocial loads and stress reactions. At the same time, it has been observed that different individuals react with widely varying physiological responses, for example, to such exposure. Consequently, not everyone reacts in the same way, even if the exposure appears superficially to be the same for everyone. We have a brain that, in a psychological sense, has encoded various experiences during our lives, and which from a biological perspective has neurons that are connected in different ways. These psychological and biological differences between individuals mean that people at work identify and deal with challenges in the most varied ways. The brain has been something of a "black box" within stress research for a long period, but where contributions from the cognitive and neurocognitive sciences have greatly contributed to a growth in knowledge in recent years. Research shows that cortisol secreted in the stress response affects certain cognitive functions (see Fact Boxes 7.3 and 7.4). The results are, however, not entirely clear cut as regards which functions are affected and what period of time that changes follow. There is, however, a great deal of evidence to suggest that the cognitive effect as a result of stress is selective, and in this way certain functions may be affected in a situation at the same time as others remain intact. In order to investigate the effects of cortisol on cognitive functions, research subjects have either been exposed to experimentally induced stress, had a pharmacologically induced rise in their cortisol levels, or already showed changes in their cortisol levels as a result of illness. Cognitive functions that have been proved to deteriorate are the declarative memory (see more on different types of memory later in this chapter) and the working memory, divided and selective attention, concentration and learning, and verbal representation. The outcomes of the various studies are not clearcut, and among the functions mentioned is also an altered cognitive capacity as a result of high cortisol levels.

7.7.4 STRESS AND COGNITIVE PERFORMANCE

The general pattern for how cognitive performance is affected by increasing exposure to corticosteroids is an inverted U-shaped function. A deficiency in stress hormones makes for poorer cognitive performance which is also produced by exaggeratedly high levels. At moderate levels of stress hormone, cognitive performance may on the other hand be improved. This inverted U was described using simpler methods (e.g., measuring the pulse) as far back as the early 1900s, and came to be called Yerkes–Dodson's Law of Motivation, which says that there is an inverted U correlation between the level of exertion and the level of performance.

7.8 WORK ORGANIZATION CONDITIONS IMPACTING ON MENTAL LOAD

If the work organization circumstances at a workplace are not good, they constitute stressors, which may lead to stress reactions of a kind described earlier in this chapter.

When the concept of psychosocial work environment is used in the context of working life, attention is drawn to the need for a holistic view of people at work. First and foremost, people are studied in relation to factors that specifically affect their work, even if the balance between work and leisure time has been studied more intensively in recent years [Nylen et al. 2007]. Below we attempt to exemplify Victoria's work, which was described in the introduction, within areas relating to the circumstances of a psychosocial nature in the organization of the work.

Organizational characteristics. This is where the organization is often described from the perspective of a matrix (project-oriented) or line organization, for instance— hierarchical or flat organizational structures, respectively. *What does the hospital's organization look like; is it hierarchical or flat or does it have some other form? What position does a newly qualified doctor like Victoria have in this hierarchy?*

Work content. The content is reflected in descriptions of workload, repetitive tasks, influence and control, cognitive or mental load. *How is Victoria able to influence her work situation? Can she change her tasks, and in that case does she need to have this confirmed by a superior? Can she stop operating if she feels tired?*

Interpersonal relationships. This relates to the degree of support from work colleagues, subordinates, and superiors at the workplace. *What does Victoria's support from management and work colleagues look like? Is this support stronger for example from the nurses than from the management?*

The temporal design of work tasks. This relates to the organization's planned working hours, work cycles, and shift work. Does Victoria work at night? *What is the planned length of time she is allowed for each patient?*

Overtime. Treated in the same way as working hours, but here in the sense of *overtime work. In Victoria's case we can define overtime work as working hours not planned by the organization; that is, Victoria has to work longer to look after acutely ill patients who quite simply cannot be handed over to a replacement.*

Economy. Are, for example, the replacement levels, salary systems and contract systems satisfactory? *Does Victoria earn a reasonable salary in relation to the various demands of her job?*

Work status. Does the work have *status and value* for the individual seen from the perspective of those around them, *that is, how is Victoria's work valued by other people?*

7.8.1   COMMON PSYCHOSOCIAL MODELS

One of the best-known models used in research into the role of psychosocial factors in health in larger populations is the so-called Demand and Control model. Its originators [Karasek and Theorell 1990], have constructed this model as a four-field diagram (Figure 7.2), in which various types of work are placed with psychological demands on the vertical scale and control (decision latitude, powers) on the horizontal

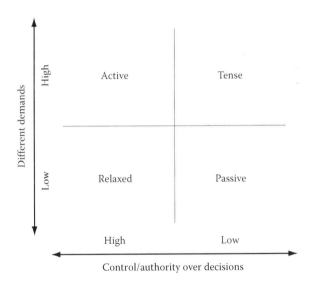

FIGURE 7.2   The Demand and Control model describes the psychosocial load at work. Reducing the demands of work is not always the best solution to avoid stress. Greater control over the work situation may be the solution. A further dimension which is sometimes to be found in the model has to do with social support, which is assumed to play an important role for how one experiences stress. (Modified from Karasek RA. and Theorell, T. 1990. *Healthy Work—Stress Productivity and the Construction of Working Life*. New York: Basic Press.)

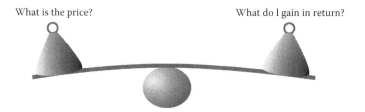

What is the price?   What do l gain in return?

FIGURE 7.3   The Effort–Reward model describes the notion that, if the individual works under great time pressure, or if the work is very demanding, for example, overtime work (effort), this must be balanced by their being treated with respect by superiors and co-workers, and receiving support in difficult situations (reward). Otherwise considerable stress is caused as a result of the lack of a reward. Illustration: Niklas Hofvander.

scale. Examples of occupations making considerable psychological demands are those with a very high work rate on assembly lines, other industrial work with a high work rate, and those involving uncertainty about continued employment and service occupations. These occupations are characterized by heavy demands in combination with low control, not only as regards the job but also as regards social and informal contacts at the workplace.

In conditions such as these, a great deal of tension is assumed to arise and to be maintained during a period when nothing at work gives any scope for activity to reduce the tension. It is assumed to continue as a "psychological or mental tension," which has consequences both for physical and mental health. A commonly occurring model which is easier to describe from an individual perspective—from the perspective of our friend Victoria, for example—is the so-called Effort–Reward model [Siegrist 1996]. Its originator, Johannes Siegrist, considers that there must be a balance between the effort demanded by the task and the reward the individual receives for carrying out that task (Figure 7.3).

The model has its origins in part in the aforementioned Demand and Control model, but focuses on reward rather than on control. It also has similarities with other previous individually oriented stress models, for example, Lazarus' contribution (see Figure 7.4). Sigesrist's model may be said to take into account Victoria's interplay between external factors in the environment (the demands and duties facing

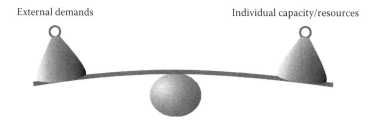

External demands   Individual capacity/resources

FIGURE 7.4   The Demand–Resource Model is often illustrated in the form of an old-fashioned pair of scales. If the individual's ability weighs light in comparison with the weight from the demands of their surroundings, or if these demands are regarded as light in relation to the individual's ability, stress results. It is best if the demands and the ability are matched, in balance. Illustration: Niklas Hofvander.

Victoria, namely demands made by the hospital and the chief consultant and the patient's needs) and internal factors in the form of the individual's ability to cope with these on the one hand (e.g., Victoria's need for control and ability to cope with stress) and, on the other, those factors comprising compensation or reward for her efforts (does Victoria know, e.g., that she receives financial rewards, recognition, or status commensurate with the demands made on her?)

A more individualistic and cognitive model that partly originates from a pioneer of stress research is Richard Lazarus' model [Lazarus 1966]. This cognitively oriented model focuses on the fact that, when the individual is faced with demands from their surroundings, the brain makes an assessment: Is this important for me? Can I cope with it? Can I use my abilities? A Swedish pioneer, Marianne Frankenhaeuser, used to express the consequences of the model in the following way: "If there is a balance between demands and one's own ability, the challenges feel stimulating and one may experience positive stress. But if the demands are too high or too low, the stress can become a torment" [Nylen et al. 2007].

7.8.2 Psychosocial Environment is Perceived in Different Ways

In order for a psychosocial demand to be perceived by the individual, it has to have some significance; that is, it must have cognitive content. For Victoria, different psychosocial demands are in many ways a more "invisible" or intangible load, and for this reason are difficult to provide with intelligible content, unlike more physical loads. When researchers measure the effects of the impact of different psychosocial factors on different individuals at the same time, we always find individual reaction patterns. One reason may be that what Victoria experiences as time pressure is not experienced in the same way by her colleagues, which is why they of course have different reaction patterns. But even if Victoria and her colleagues were to experience time pressure in exactly the same way, they react in very different ways physically and mentally, which we described in a physiological sense earlier in this chapter. For the time being, we may note that, as Victoria has learnt from previous experiences and as these experiences differ from those of her colleagues, she remembers, behaves, and reacts differently to psychosocial challenges. The psychosocial challenges have, quite simply, different cognitive content for Victoria and for her colleagues. Nor are we equal in a biological science (e.g., the neurons in our brains are not connected in the same way). The physically delimited workplace is therefore experienced differently by different people. For example, not all individuals at a physically delimited workplace find the work interesting or experience stress. This means that a workplace in the singular does not exist in any sense if we ask more than one individual about their workplace.

7.9 COGNITIVE LOAD

Biologically different cognitive loads (see the previous section on the SAM system and HPA axis, for example) may imply stresses that alter or threaten to alter the "internal environment" of the body, for example, in the form of a hormonal or endocrinal change or as a physiological load on the processes within the central nervous system (CNS). Psychologically one can see a load within the cognitive systems, which can be measured in the form of a measurement of performance, concentration, involvement, and different kinds of memory capacities.

It is on the whole difficult to imagine any job which does not involve both physical and cognitive/mental loads and challenges. The slaughterhouse worker (Chapter 5) has a job characterized by manual handling involving major muscle groups such as the muscles of the torso and legs. Physical load may result in considerable strains on his musculoskeletal system; for example, on his muscles, joints, tendons, and ligaments. But working in a slaughterhouse also implies a cognitive load, as the demands for precision and perhaps also the feeling of being under time pressure may be pronounced. Tasks demanding both precision and speed are jobs that imply that mistakes may occur. In the case of the butcher, this would perhaps be in the form of cuts if his knife slips. Just as the physical load varies, the cognitive load in all probability changes, for example, depending on what part of the animal is being butchered. The greater the precision required, the more cognitive resources have to be allocated, and the more refined the motor feedback required for the task. From a cognitive perspective, the brain works in different ways depending on whether what the butcher is working on is connected with learning something new or whether the action is *automated* (routine tasks). If the action is strongly automated, the work as a rule is carried out at what memory research calls a *procedural memory level* (read more about procedural memory level in the next section). Something that, on the other hand, is quite new or unexpected requires greater mental effort and closer attention to be able to carry out the work. In order to be able to do two things simultaneously, it is necessary that these are automated, so that neither of them requires one's total attention. This is not least important in occupations where the risk of injury is high. As the *working memory*, which could be said to be a measure of the processing capacity of our thinking, is limited, there is not a great deal over for anything else when the butcher devotes himself to what is not automated. A high allocation of cognitive resources is, for example, found in a job constantly requiring the individual to have a very limited opportunity of thinking of anything other than what must be carried out at that precise moment.

7.10 ROUTINE WORK IS CARRIED OUT AT THE LEVEL OF PROCEDURAL MEMORY

All work, now as in earlier times, involves forms of mental or cognitive load. Care professions (Chapter 3), the work of the electrician (Chapter 4), and the building worker (Chapter 9), working on computers (Chapter 6), the work of the long-distance lorry driver and bicycle messenger (Chapters 2 and 8), all challenge cognitive systems in different ways and different tasks at different moments, for example, one work cycle. If there is no danger (e.g., icy roads) present, it could be said that an experienced bicycle messenger or driver is working on an automated level, or procedural memory level (read more about procedural memory in the next section). Characteristic of this level is that the work can be carried out at the same time as it is possible to think about something else. That is to say, an experienced bicycle messenger at work, when the weather is good and there is little traffic, can be thinking

TABLE 7.1

Examples of Sources of Information to which a Bus Driver Needs Access in Order to Control his Vehicle

Sources of bodily information—*that is, sources of the kind that do not require conscious assessment, or require very little conscious assessment.*	Cognitive sources of cognitive information—*that is, sources of the kind requiring more advanced assessment.*
	After it is memorized, this information is retained at a procedural level, but rapidly climbs to higher levels and processing, for example, in
Sensory skin input	*case of danger.*
Motor feedback	
Balance sensations related to movement of the bus	Status of various instruments
Fatigue/wakefulness	Situation of the passengers
	Noise/vibrations
	Light conditions
	Position of the bus
	GPS
	Relative speed
	Fuel
	Surroundings
	Geography
	Weather/road conditions
	Traffic density
	Time

about things completely different from his work. But even at this level it is necessary for a driver, for example, to use many different sources of information to be able to drive his vehicle safely. In Table 7.1, examples are provided of various sources of information of this kind.

As soon as a cognitive information source indicates a deviation from the normal (e.g., the road surface suddenly becomes icy), the automated level is abandoned, and cognitive structures dealing with more complex information take over; the cognitive load increases. It may be similar for the doctor, Victoria. An operation which is almost routine suddenly becomes dramatic as a result of unexpected internal bleeding in the patient. This increase in cognitive load also launches our stress systems in the way described previously.

7.10.1 Organization of Work from a Cognitive Perspective

Changes in production processes can be described from a cognitive perspective. New forms of production imply that anyone who has worked and thought in a particular way will have to think in a different way in a new form of production. The "Ford" assembly line used in the car industry (sometimes a little inaccurately called the Taylor line), with its relatively simple and short physically executed work cycles, was considered by many people to be belittling, as the individual was regarded like a robot who could be replaced when worn out. The operations were considered to be so easy to carry out that no mental activity seemed to be needed, that is to say, thought and action could be separated. The work was carried out at the procedural memory level. During the 1970s, many of these traditional assembly lines were replaced by autonomous work teams with job enlargement in the form of larger units, greater responsibility for quality, and longer work cycles. This method of working made quite different cognitive demands and resulted in different mental loads from the more traditional assembly line. Individual people had to plan their work in a different way from previously; planning needs the ability to think ahead—that is, to make use of what within memory cognition research is called *prospective memory*. Prospective memory has to do with the fact that that what has to be remembered occurs in the future, for example, remembering to submit the stores report before a specific date. Prospective memory can help us to reduce stress, as with the aid of planning for the future we can prevent what is worrying us from becoming as serious as it would have been without any planning. On the other hand, in situations in which we feel more helpless, for example, the threat of losing our jobs, prospective memory helps create anxiety. Other cognitive abilities or memories also need to be used in an increasingly complex process and these affect, for example, the *declarative* memory, which helps us to verbalize, for instance, complex processes so that all learning does not need to be done completely from scratch, and shortcuts for how the work is to be carried out can be transferred between experienced and inexperienced co-workers. In the increasingly complex environment greater demands are also made on *semantic* and *episodic* memories, which both, roughly speaking, have to do with learning and remembering specific and more general factual information about the work process [Melin 2003].

Traditional assembly line work in this way became more cognitively complex and thought and action became more intertwined. The more intertwined the higher cognitive levels were with the action itself, the higher the cognitive involvement and less scope to think about anything other than a single consequence. It is a good thing if the work content leading to the cognitive involvement is felt to be interesting, meaningful, and contributes to development. It is not as good if the work content lead to cognitive involvement perceived as uninteresting or monotonously repetitive. Then we can talk of a mental or cognitive strain rather than work which is cognitively stimulating. An information flow leading to a constant cognitive involvement results in the individual having a very limited opportunity for thinking of anything other than what has to be carried out at that moment. If this information flow is felt to be monotonously uninteresting, we have a work situation that is difficult to manage and which in the long term becomes a cognitive strain. One question is whether jobs today contain "more" cognitive strains of the kind that is described above. From this perspective the increased establishment of call centres, for instance, is interesting (Chapter 6).

The ability of using one's senses to absorb and manage technically transmitted information mentally or cognitively has become an increasingly important part of most occupations today. This is the case not least within nursing care, with the introduction of electronic records and referral management, appointment systems, and computerized financial accounting systems. At the same time as technical systems are becoming ever more common within occupations involving people contact, another trend in industry may be noted, where one sees increasing elements of social demands. One study [Melin et al. 1999] shows that, at industrial workplaces where earlier research has been taken on board, and among other things modern job rotation has been introduced with longer and more varied work cycles, this leads to greater demands for social interaction and cooperation. By rearranging the work at a car assembly plant according to new research recommendations, the workers became less tired during their shift and had lower pulse rates and fewer stress hormones in their blood. In addition, the car fitters could relax better after work and recover more quickly. There was only one thing that the workers thought needed extra effort, and that was on the social plane. They were forced to interact more with their workmates in order to coordinate the work, which could sometimes lead to minor conflicts.

These observations agree well with the trends that many people can see, namely that the mental strain is increased but in a different way, within the care professions and in industrial occupations. This applies, among others, to mental demands as a result of new technical systems in nursing, for example, and—within industry— greater mental demands as a result of the demands for social competence. These greater demands for social ability apply not merely to "workers of the future," who move between short project jobs and are dependent on well-developed networks, but may therefore also come to affect employees in traditional manual work.

The psychosocial environment, and thereby the cognitive strains, have undergone major change over time, while the stress reactions remain the same. This applies both to assembly lines and nursing. The doctor, Victoria, is presumably cognitively challenged and has mental strains in a different way from those of her colleagues just a few years ago. The mental load originates, on the one hand, from the work with patients itself, and on the other, to a large extent from comprehensive organizational changes and new IT-related administrative operating systems that have to be used by the doctors themselves.

7.10.2 Link between Behaviour and Cognitive Load

Within research there is no conceptual agreement as to what is considered to be mental or cognitive load. Concepts such as *mental variation* occur frequently in connection with repetitive work in an industrial environment. Many industrial occupations involving assembly and packaging may be characterized as repetitive and monotonous, which are factors increasing the risk of developing musculoskeletal problems, in particular in the upper extremities. For this reason, an increase in variation in the work, both physical and mental, is often suggested in connection with the process of change within the industry.

Job enlargement, for example, job rotation, a different pattern of breaks, and an incentive to interrupt repetitive movements, are frequent proposals within this type of industrial occupation so as in this way to achieve greater variation. In other working environments, for example, those that are strongly computerized such as work in control rooms, but also many service occupations that are not primarily regarded as physically demanding, the concept of cognitive load is used without direct reference to variation. What might be meant by cognitive load is, for example, that the computer users have to exert themselves mentally, for example, in order to retain the instructions in their memory. A high cognitive load arises, for instance, if the system does not provide support to the user, and the users themselves have to remember the elements. In order to avoid this, it is usual to suggest, for example, having drop-down menus where the user's various alternatives are visible. Here the point of reference is the encounter between the different mental models of the system designer and the user. In other words, the user has a mental model of the system that has been created according to the user's own tasks, previous experience, documentation, education, what the system looks like and how it behaves. On the other hand, the system designer also has a mental model, but this is created on the basis of how the system is constructed, for example, with different procedures, functions, access routes or database structures, and also on the basis of what the system looks like graphically. The designer's model and the user's model do not correspond, and a discrepancy arises, which is why many cognitive resources have to be allocated on the part of the user to adapt to the system.

7.10.3   LINK BETWEEN COGNITION AND EMOTION

Distinguishing cognition from emotion is a classic problem, for which there is no space for discussion within the framework of this chapter. The finely calibrated facial expressions of the human face are unique to us. In many species, as in human beings, a common emotional expression is, for example, to shrug the shoulders in cases of threat. Birds raise their wings and fur-bearing animals bristle. It should be emphasized that cognitive stress is rarely disconnected from the emotional features of our everyday lives. Often cognitive stress (e.g., the feeling of being under time pressure) is used as if the phenomenon was free of emotional charge. The experience of time pressure, for example, is unique to a being that has the cognitive ability to comprehend time. Human beings have the ability to understand clock time, and therefore a purely cognitive ability. For Victoria, time pressure is something more than experiencing the difficulty of finding time for something vital, maybe a patient who needs care—and it thereby becomes an emotionally loaded stressor. Time pressure of work can therefore not merely be regarded as a clearly cognitive stress, but also as a strongly emotionally charged stressor, which is why there are reasons to assume that as soon as we talk about cognitive or mental load, it is also a question of emotional stress and not merely cognitive stress.

7.11   MENTAL LOAD: EFFECTS ON THE MUSCULOSKELETAL SYSTEM

The American physiologist and pioneer in the field of stress research, Walter B. Cannon, wrote at the beginning of the last century that muscle tension in the body is an appropriate reaction when we are subject to stress and threats. This muscle reaction makes possible a greater readiness to cope with impending danger. As early as the 1930s, some experimental studies were made on human beings that, with the methodology of that time, studied muscle activity in relation to mental stress. Among others, so-called focus attention studies were carried out, which for instance mean that individuals at rest had to imagine themselves lifting various objects. Greater muscle activity could be registered in the muscles relating to the lifts the individual was imagining. In a study of this kind a kind of dose–response correlation was found; the heavier the object the individual imagined that they were lifting, the greater the muscle activity. Based on a hypothesis that deaf people, who used sign language to a greater extent than those who could hear, would show greater activity in the muscles of the hand also during "non-verbal" problem-solving tasks, it was discovered that this was indeed the case, using the most modern methods of the time.

7.11.1 Mental Load and Muscle Activity

As work generally always involves a certain measure of physically induced muscle activity, it is interesting to ask whether mental stress contributes to greater activity in the trapezius apart from the pre-existing physical load [Melin and Lundberg 1997; Lundberg 2002]. Several studies have been carried out to investigate this. An example of studies of this kind is a relatively early one in which an attempt was made to keep the physical load constant during two different experimental situations while a stressful mental arithmetic task was imposed in one of the situations. The constant physical load consisted of the subjects having to move not particularly heavy balls between two containers. Muscle activity monitored using EMG (see Chapter 6, Section 6.12.6) showed a very strong increase in activity in the trapezius (75% increase) in the combination of moving the balls and doing mental arithmetic compared with merely moving the balls. A similar increase in activity, though not as strong, has been measured in other studies. Among other things, training sessions and competitive sessions have been compared for pistol-shooters. Here the weight of the pistol comprises the constant physical load and the element of competition the mental load, while the training session involves less mental stress. The training session generated greater activity in the trapezius both just before the competition session and during the training session. Similar patterns have been found in female checkout staff, which showed that mental stress combined with physical stress resulted in greater muscle activity than the presence of physical stress or mental stress alone. In some studies varying degrees of difficulty at work have been studied in work at computer terminals, where it is possible to see a pattern characterized as follows: the higher the degree of difficulty, the higher the muscle activity. A common way of studying the degree of cognitive complexity is to use what is called the Stroop test. This test provides conflicting information that generates stress in the test subjects. In several studies the test has been shown to lead to increased EMG activity beyond what has already been produced by relatively low physical load. In a Danish study, among others, the Stroop test was used in conjunction with work on a computer. The researchers found greater muscle activity in the trapezius both when using the mouse and the keyboard.

In other studies the degree of cognitive/mental complexity has been varied. In one experiment the complexity was varied between two levels (low and high complexity) in conjunction with a reaction time test on a computer monitor. Here it was found that, generally speaking, all the test subjects reacted with a somewhat raised EMG activity under both sets of conditions. When the degrees of complexity were compared, it was found that eight out of the 18 test subjects in repeated provocations consistently generated a higher EMG activity in the trapezius at a high degree of complexity compared with activity at a low degree of complexity, while other subjects showed a more varied pattern. An experimental study of female checkout staff in a department store showed that cognitive load combined with physical stress resulted in greater EMG activity than the presence of physical stress or mental stress alone.

Emotion has been studied in the form of anxiety when faced with an unprepared oral presentation on the part of 42 male university students. The study was intended to investigate physiological reactions during and after preparations and to compare extroverted with introverted individuals. One of the physiological measurements was surface EMG activity in the trapezius. Introverts increased their muscle activity more during preparations than extroverts. Introverts also found it more difficult to reduce their muscle activity during the stipulated rest that followed the preparations.

There are also studies showing no effects of a cognitive load above the level to which the physical load alone contributed. Presumably, the cognitive load does not contribute to any further measurable muscle tension if the physical load is high. How mental load and light physical loads may contribute to disturbance and pain in the muscles is shown in Chapter 6.

7.12 PHYSIOLOGICAL MEASURES OF MENTAL LOAD

Today there are different brain imaging technologies that provide a great deal of information about how various brain structures are involved in the stress process.

7.12.1 BRAIN IMAGING TECHNOLOGIES

Magnetic resonance tomography (MRT) works in such a way that the resonance imaging camera causes the hydrogen nuclei in the body to change their magnetization with the help of a combination of magnetic fields and radio waves (RF pulses). After each RF pulse the atomic nuclei revert to their original magnetization, emitting radio waves at the same time. The radio waves are picked up by an antenna, and a computer converts the information into a series of cross-sectional pictures. An example is studies of the differences in oxygen content arising in the brain during a certain challenge/function. For example, the MR signal in the motor centre of the brain changes when a finger is moved. This is called functional magnetic resonance imaging.

Positron emission tomography (PET) studies are used primarily to discover and locate tumors. The technology has also been used in stress research. This technology measures, among other things, blood flow and the use of glucose in various organs, primarily in the brain and heart.

7.12.2 BRAIN ACTIVITY

Brain activity is a good measure for studying cognitive demands at work in certain restricted situations. The electroencephalogram (EEG)—that is, a representation of the electrical activity of the brain—is often used as a measure of brain work in cognitive tasks. The EEG spectrum is analysed to determine how high the activity is when performing various cognitive tasks. By recording activity at the same time using electrodes at several points on the crown of the head, one can produce a topographical map showing the distribution of electrical activity across the crown. Examining these topographical maps shows which parts of the brain are active during different periods of a complex work process under high mental work load. What is primarily studied is how the effect changes in the so-called alpha and delta bands. The others are the theta and beta bands. Mental work loads reduce the effect in the alpha band and increase the effect in the delta band. EEG as a method is very sensitive to blinking and head movements. These movements therefore have to be measured and, by using computer programmes, the effects to which they give rise have to be filtered out.

7.13   MORE INDIRECT METHODS

Cognitive load can, as has previously been described, be seen as a strain on the processes in the CNS. For this reason, there are possibilities that the load on the CNS also affects physical processes other than those directly linked to information processing in the system. This gives us opportunities to measure indirectly the cognitive work load with the aid of physiological indicators. Some examples are given below.

7.13.1   HEART RATE

HR or pulse is a commonly used indicator of mental workload. When the mental load increases, the pulse also increases. The two most common methods of measuring pulse are HR, which is quite simply the number of heart beats per minute, and Interbeat Interval (IBI), where the time between beats is measured. The monitoring equipment required is relatively simple, but one has to be aware that the very simplest equipment sometimes measures the pulse as an average over too long a period, which can lead to the loss of important information. One method of acquiring good data is to measure IBI continually with EKG and subsequently convert to HR. Then we get both a high degree of accuracy and a measure which is easy to relate to HR. As the main task of the heart is to provide the body with blood, there is a risk that results become misleading if major changes occur in the physical load during or between tests. One can therefore advantageously supplement HR with methods that measure movement, for example, with actigraphic monitoring (see also Chapter 6, Section 6.12). If there is a high correlation between, for example, body movement measured using actigraphy and pulse, it is probable that an increase in pulse will result from the movements in themselves and not from the mental load. A simple method for monitoring HR is to use the so-called pulse clocks (Chapter 6, Section 6.12).

7.13.2   HEART RATE VARIABILITY

Heart rate variability, HRV, has been used as a measure of mental work load. What we mean by HRV is the periodicity by which the HR varies. Normally, this is five to ten times a minute. High cognitive activity is assumed to reduce HRV.

Some researchers are doubtful about using HRV in complex cognitive tasks. They consider that such tasks reduce HRV so much that it reaches a level where it cannot decrease any more, and thereby no longer functions as a measure of load. Other researchers consider that the use of the technique for monitoring cognitive load should be avoided, as research indicates that it measures emotional rather than cognitive work load.

7.13.3 SWEAT GLAND ACTIVITY

The sweat glands are found in large numbers in the palms of the hands and on the bottom of the feet. Galvanic skin response is used to measure sweat gland reaction activity in the skin. The skin's conductivity changes when people react—consciously or unconsciously—to mental load and emotions. The emotion that arises, even before it reaches the conscious stage, creates a reaction in the skin. The method therefore registers changes very quickly. In the conscious phase we often notice ourselves sweating.

7.13.4 BLINK RATE

Blink rate is a common measure in research into mental load. Blink rate tends to decline when visual demands increase. This is a result of being practically blind during the period that the eyelid is closed, which leads to the possibility of discovering important visual events increasing as blink rate decreases.

7.13.5 PUPIL DILATION

Research has shown that the pupil increases in diameter when mental load increases.

7.13.6 RANDOM EYE MOVEMENTS (ENTROPY)

When mental workload increases, random eye movements tend to decrease. This is measured by equipment that registers where one is looking and the fact that one then analyzes the pattern one sees.

7.13.7 BREATHING

Breathing involves several parameters that can be used to monitor mental workload. Airflow, air volume, temperature, and variation in inhaling and exhaling can be monitored. As the monitoring is disturbed by speech, its use is limited in many contexts. The method is used in such things as studying cognitive load in connection with computer work. The gas exchange between the lungs and the blood is also studied.

7.13.8 ELECTROMYOGRAPHY

See Chapter 6, Section 6.12.6.

7.13.9   Hormones

As described earlier in the chapter, the monitoring of stress hormones is frequently carried out in connection with research-related stress studies (see the descriptions of SAM and the HPA axis earlier in the chapter).

7.14   SUBJECTIVE MEASURES OF MENTAL LOAD

Subjective methods have proved to be particularly relevant for monitoring mental workload, but there is no method that works by itself in all situations [Gawron 2000]. Some methods are very similar to each other, others are very different. Some use broader questionnaires, whereas others use more specific scales for monitoring specific work situations. Examples of the latter are: *The Cooper-Harper Rating Scale*, the *Bedford Workload Scale*, the *Overall Workload Scale*, the *NASA TLX*, and the *SWAT* [Hancock and Meshkati 1988].

7.14.1   Cooper–Harper Rating Scale

The Cooper–Harper scale has the form of a tree diagram. It was created in 1969 by George Cooper and Bob Harper, two test pilots from the United States. The intention was to create a method for evaluating new or modified (primarily military) aircraft as regards how easy the plane was to handle in different situations. It also provides a measure of how great a mental load the pilot is experiencing. The Cooper–Harper scale is divided into 10 steps, where one represents excellent properties (very low mental workload) and 10 represents major deficiencies (overload). The Cooper–Harper scale is something of a standard scale for evaluating the handling characteristics of an aircraft in different situations. It measures both performance and mental workload and is relatively simple to use.

7.14.2   Modified Cooper—Harper Rating Scale and Bedford Workload Scale

The Modified Cooper–Harper Scale (MCH) was developed to be used in those cases where the task was not primarily motor or psychomotor. A scale was needed that better captured cognitive functions such as, for example, perception, observation, evaluation, communication, and problem-solving. In brief, the following changes were introduced: the terminology was modified in the tree diagram; the endpoints of the scale were changed to "very easy" and "impossible," and the pilots were asked to assess the mental workload rather than the effort required to control the aircraft. The Bedford Workload Scale is a modification of the original Cooper–Harper scale. The structure is similar to the Cooper–Harper scale with a tree structure, but the terminology has been changed. The individual has to decide whether the task could be completed, whether the mental workload was acceptable, or whether it was satisfactory without a reduction of the load. The scale uses an assessed reserve capacity on the part of the pilot during different situations to assess their mental workload.

7.14.3 Overall Workload Scale

The Overall Workload (OW) Scale is a bipolar scale (ranging between "very low" and "very high" from left to right), on which test subjects estimate their mental load over 20 steps.

7.14.4 NASA TLX

In the NASA Task Load Index (TLX) six dimensions are used to measure workload: mental demand, physical demand, time pressure, how well the task was carried out (performance), exertion, and frustration. The concept of workload is used here instead of mental workload. This is because NASA-TLX is a so-called multidimensional method, which attempts to capture more aspects of the concept of workload—physical demands, for example.

7.14.5 SWAT

SWAT (Subjective Workload Assessment Technique) is, like NASA-TLX, a multidimensional scale. In trials with SWAT, three scales were used with three steps in each. The scales are: (1) time pressure, which reflects the time available for planning, carrying out and supervising a project; (2) mental load, which assesses how much conscious mental effort is required to carry out a task; and (3) psychological stress load, which estimates how much confusion, frustration, risk, and anxiety is associated with the trial.

7.14.6 Broader Subjective Instruments

Then there are a number of instruments to monitor stress in a wider perspective. Earlier in the chapter mention was made of Karasek's and Theorell's model, and Siegrist's and Frankenhaeuser's models. There are variants of the questionnaires and scales for these models. One might add Spielberg's and Vagg's "Job Stress Survey," which is also to be found in several languages, and which aims to identify the sources of work-related stress at the workplace [Spielberg and Vagg 1999]. The instrument is answered first as regards how seriously various stressors are felt to be, and subsequently a decision is made on how often they occur. In this way, information is acquired about the degree of seriousness and frequency of the stressors.

7.15 WHAT DOES THE LAW SAY ABOUT MENTAL LOAD?

All the member states in the EU are covered by a general directive which applies to safety and health at work [i] (Directive 89/391/EEC). This also includes mental health at work. When this applies to working conditions in computer work, there is also a minimum directive with official communications affecting mental load [ii] (Directive 90/270/EEC).

As regards legislation in the United States, see Chapter 10.

7.16 SUMMARY

Allostasis means the process that ensures that the body's physiological systems are in balance, and that homeostasis is thereby maintained during varying external conditions. Allostasis has a primarily protective function for the body, but allostatic processes can also act to the detriment of the body and produce harmful effects if they do not disappear within a reasonable period. That is, they can give rise to *allostatic load.* It is therefore important to be able to recuperate, irrespective of which load one is subject to. In this chapter we have said that, as regards exposure, it may seem self-evident to distinguish between physical and mental stressors. The physical stressors have an impact on the internal environment of the body without higher cognitive functions needing to be employed. As regards the stress reaction itself, it is interesting that the division into physical and mental stress reactions is not as obvious. We have also said that cognitive functions may be disturbed as a result of prolonged mental stress, and cortisol (the HPA axis) is suspected of playing an important role in this context. In this chapter we have described how different memories such as, for example, the work memory and procedural memory, are made used in different operations. Routine work is often done at the procedural memory level, while newer tasks and those requiring more problem-solving skills require a different kind of memory use. In this chapter we have also mentioned different methods of monitoring mental load, for example, direct brain imaging methods to more indirect physiological measurements and self-reported feedback. The introduction of new technology has brought with it new cognitive demands in almost all areas of working life. New technology in many cases implies radically new working methods, and challenges that are not always entirely easy to discover, as mental exposure is more inaccessible (more invisible) than physical exposure. All member states in the EU are covered by a general framework directive applying to safety and health at work including mental health at work.

REFERENCES

Alderson, AL. and Novack, TA. 2002. Neurophysiological and clinical aspects of glucocorticoids and memory: A review. *J Clin Experim Neuropsychol.* 24(3):335–355.

Directive 89/391/EEC—*On the introduction of measures to encourage improvements in the safety and health of workers at work.* European Agency for Safety and Health at Work. http://osha.europa.eu/en/legislation/directives/the-osh-framework-directive/

Directive 90/270/EEC—*On the minimum safety and health requirements for work with display screen equipment.* European Agency for Safety and Health at Work. http://osha.europa.eu/en/legislation/directives/the-osh-framework-directive/

De Kloet, ER., Vreugdenhil, E., Oitzl, MS., Joels, M. 1998. Brain corticosteroid receptor balance in health and disease. *Endocr Rev.* 19(3):269–301.

Eurostat. 2008. National Accounts, annual average. http://ec.europa.eu/publications/booklets/eu_glance/66/index_en.htm

European Agency for Safety and Health at Work, Bilbao. 2004. http://europa.eu/legislation_summaries/employment_and_social_policy/health_hygiene_safety_at_work/c11110_en.htm

Gawron, VJ. 2000. *Handbook of Human Performance Measures.* Mahwah, NJ: Lawrence Erlbaum Associates.

Gilbertson, MW., Shenton, ME., et.al. 2002. Smaller hippocampal volume predicts pathologic vulnerability to psychological trauma. *Nature Neuroscience* 5:1242–1247.

Gurvits, TV., Shenton, ME., et al. 1996. Magnetic resonance imaging study of hippocampal volume in chronic, combat-related posttraumatic stress disorder. *Biological Psychiatry* 40:1091–1099.

Hancock, PA. and Meshkati N. (eds.) 1988. *Human Mental Workload.* Netherlands: Elsevier Science Publishers B.V.

Karasek RA. and Theorell, T. 1990. *Healthy Work—Stress Productivity and the Construction of Working Life.* New York, NY: Basic Press.

Kristenson, M., Garvin, P., Lundberg, U. (eds.) 2011. *The Role of Saliva Cortisol Measurement in Health and Disease.* Bussum, Holland: Bentham Science Publishers. Open Access.

Lazarus, RS. 1966. *Psychological Stress and the Coping Process.* New York, NY: McGraw-Hill.

Lundberg, U. 2002. Psychophysiology of work: Stress, gender, endocrine response, and work-related upper extremity disorders. *Am J Ind Med* 41(5):383–392.

McEwen, BS. 2000a. The neurobiology of stress: From serendipity to clinical relevance. *Brain Res* 15:172–189.

McEwen, BS. 2000b. Allostasis, allostatic load, and the ageing nervous system: Role of excitatory amino acids and excitotoxicity. *Neurochem Res* 25(9–10):1219–1231.

McEwen, BS. 2006. Protective and damaging effects of stress mediators: Central role of the brain. *Dialogues Clin Neurosci* 8(4):367–381.

Melin, B. 2003. "Mental assembly lines" risks for cognitive overload. In C. Otter von (ed.) *In and Out in the Swedish Working Life* (page 235–251). Stockholm: NIWL (In Swedish).

Melin, B., Lundberg, U., Söderlund, J., Granqvist, M. 1999. Psychological and physiological stress reactions of male and female assembly workers: a comparison between two different forms of work organization. *J Organiz Behavior* 20:47–61.

Melin, B. and Lundberg, U. 1997. A biopsychosocial approach to work-stress and musculoskeletal disorders. *J Psychophysiol* 11:238–247.

Nordling, S. 2003. *Stress and Exhaution within a Neuropsychological Perspective, An Overview.* Inst for Cognitive Science, Umeå University. (In Swedish).

Nylén, L., Melin, B., Laflamme, L. 2007. Interference between work and outside work demands relative to health, unwinding possibilities among full-time and part-time employees. *Int J Beh Med* 14:229–236.

Ohlin, E. 2001. Studies of physicians' work-situation. *Physicians Magazine* 38(101):2865–2866. (In Swedish).

Sapolsky, RM. 2003. *Why Zebras Don't get Ulcers.* New York, NY: Henry Holt and Company.

Siegrist, J. 1996. Adverse health effects of high effort/low reward conditions. *J Occup Health Psychol* 1(1):27–41.

Spielberg, CD. and Vagg, PH. 1999. *Job Stress Survey JSS.* Odessa: Psychological Assessment Resources Inx.

FURTHER READING

European Journal of Applied Physiology (EJAP) 2006 (2) has a special edition on mental and physical load and can be found as a PDF at the following link: http://www.springerlink.com/content/1439–6327/ The same journal also has a previous and important special edition in this field with an editorial signed by Gisela Sörgaard, Ulf Lundberg, and Roland Kadefors.

Linton, S. (ed). 2000. *New Avenues for the Prevention of Chronic Musculoskeletal Pain and Disability.* Amsterdam, New York, NY: Elsevier Science.

Nilsson, LG., Adolfsson, R., Bäckman, L., de Frias, CM, Molander, B., Nyberg, L. 2004. Betula: A prospective cohort study on memory, health and ageing. *Ageing Memory and Cognition* 11:132–148.

Sapolsky, RM. 2003. *Why Zebras Don't Get Ulcers.* New York, NY: Henry Holt and Company.

Below are some references to get advice on how to design one's own questionnaire for specific purposes, for example, in working life.

McColl, E., Jacoby, A., Thomas, L., et al. 2001. Design and use of questionnaires: A review of best practice applicable to surveys of health service staff and patients. *Health Technol Assess* 5:1–256. Comprehensive systematic overview of questionnaire methodology. Deals with, for example, layout, phrasing of questions, response format, administrative methods and drop-out. Freely accessible on—www.ncchta.org or through ELIN.

Scientific Advisory Committee of the Medical Outcomes Trust. 2002. Assessing health status and quality-of-life instruments: Attributes and review criteria. *Quality of Life Research* 11:193–205. Detailed criteria for assessing the quality of monitoring instruments, including a discussion of explanations. Freely available via ELIN.

8 Work That Disrupts the Diurnal Rhythm

Torbjörn Åkerstedt

Photo: Joakim Romson

CONTENTS

The time is around 7 am; George, 57, is getting out of bed and thinking about the day ahead—and the evening's night drive. In the morning he makes sure that the children get to school, does the shopping, and cuts the lawn. In the afternoon he tries to take a nap. He gets just half-an-hour's sleep before it is time to drive to the haulage contractors at 5 pm. He collects his brand-new 25 m-long truck with its 620 hp V8 diesel engine, which conforms to the EU's new exhaust gas directive, and drives to the freight terminal. Then he begins a couple of hours work, loading the truck and doing paperwork. At 8.15 pm he sets off. There is only moderate traffic and driving is easy. Towards midnight, darkness and mild fatigue creep over him and it is time for a break. He has a late dinner at a roadhouse, drinks coffee, buys a couple of bottles of Coca Cola, and continues on his way. For a while he thinks about his eating habits, his weight problem, and the company doctor's warning about the risk of diabetes—his blood values are not ideal. What to do?

The road is now relatively empty and soporific, and the effects of the coffee only last for about an hour. The fatigue becomes obvious, and his stomach makes itself felt. George turns up the volume on the CD player, winds down the window, drinks a Coca-Cola, and tries to get into touch with his fellow drivers on the CB. This helps a little against the fatigue. At 4 in the morning, the fatigue is almost overpowering, but now he is approaching his destination and he is not tempted to take a further break, even if his eyelids are heavy. He wonders whether he nodded off briefly just now. But there is not much else to do other than continue on—there are no stopping places for heavy vehicles in the area. Luckily he succeeded in taking a nap before he set off, and so it should be possible to drive on for another hour. He arrives at his destination as the sun is rising at 5 am.

There he hands over his truck to a colleague who will be taking it further on. Slightly groggy from fatigue, he goes into the overnight room above the garage and lies down on the bed exhausted. He sleeps for about 5½ h and is woken at around 11 am by a combination of his own biological clock and noise from the terminal. He feels that he has slept too little, but it is not possible to get any more sleep just now. He will have to try to take a nap later, before he takes a new trailer back the next night.

8.1 FOCUS AND DELIMITATION

It is not only George's occupational group that has experiences of this kind. They also apply to pilots, locomotive drivers, boat crews, workers in the process industry, doctors, nurses, and even economists and technicians in global companies that never close. Most of these people experience fatigue and sleep disturbances. Some of them also experience stomach reactions. Many of them find themselves in hazardous work situations. Drivers are, however, the group in which impaired performance resulting from sleepiness or fatigue has the most obvious consequences. Below, we have tried to summarize what working hours that disturb the diurnal rhythm look like, what the law has to say, what the health effects are, what the physiological reactions look like, what the effects are of different types of working hours, and what countermeasures could possibly be introduced.

In this chapter, the following issues will be dealt with:

- Is George's sleepiness a safety risk?
- Can people drive a heavy vehicle at any time and for any length of time?
- How much sleep do they have to have?
- For how long can they be awake?
- Why do we sleep?
- How do we prevent dangerous fatigue?
- How effective is a nap?
- Is health affected by shift work?

8.2 DEFINITIONS AND SCOPE

What diurnal rhythm-disrupting work time systems are we dealing with? If day work involves the hours between 7 am and 6 pm, most other working times may disrupt the diurnal rhythm, particularly if night work is involved. A combination of morning, afternoon, and night shifts (three-shift work) is often used (see Figure 8.1), especially in traditional industrial work. In Europe, the employee often alternates between these shifts, whereas in the United States permanent shifts are more common. If night work is not required, the resulting schedule is often some variety of two-shift

Team	Mon	Tue	Wed	Thu	Fri	Sat	Sun
Slow rotation							
1	M	M	M	M	M	M	M
2	–	A	A	A	A	A	A
3	A	–	–	N	N	N	N
4	N	N	N	–	–	–	–
5	–	–	–	–	–	–	–
Rapid rotation							
1	M	M	A	A	–	–	N
2	N	N	–	–	M	M	A
3	A	A	–	–	N	N	–
4	–	–	M	M	A	A	–
5	–	–	N	N	–	–	M

FIGURE 8.1 Examples of shift schemes. Figures indicate week and shift team. A = afternoon shift, M = morning shift, N = night shift, – = day off.

work (morning and evening shifts). The term "shift work" is here used in a generic sense. There are a number of other work schedules that do not divide the 24 h period into equal portions, but are more irregular. This is characteristic of health care and transport work, for example.

A new form of working hours is the *12 h shift*, which often means that the work alternates between a day shift and night shift. As a worker then clocks up a 36 h week in 3 days, this provides more days off—4 days every week. This has become particularly popular in Europe by shift workers who make good use of the free time—workers who: moonlight, study, like the outdoors, do sports, and other groups.

On-call work means that the employee is available for a largely immediate work input over a certain time period. The amount of work is expected to be considerably less than what is usual during a normal shift. The proportion of work is, however, unspecified.

8.3   PREVALENCE OF UNSOCIAL WORKING HOURS

According to Eurofound (www.eurofound.eu), ~20% of those employed in the EU work "shift work." About one-third of these people at least occasionally work night shifts. The definition of "shift work" is unspecified, but probably refers to employees who do not have regular working hours each day.

8.4   WHAT THE LAW HAS TO SAY ABOUT WORKING HOURS

European working hours are regulated through the Working Time Directive (Directive 2003/88/EEC). This does not include a ban on night work, but on the other hand it does have rules about a maximum 8 h shift (on average) for night work, a ban on overtime in connection with night shifts, a ban on shifts longer than 8 h in connection with particularly sensitive work, and directed medical examinations of night-shift workers (permanent night work—not shift work). Furthermore, the shortest daily rest period is set at 11 h, the maximum working week at 48 h on average (including overtime) and the shortest weekly rest period at 35 h (including a preceding 11 h daily rest period). There can be a deviation from the directive if compensation is provided in the form of time off later. Recently, the law has been modified so that previously "exempt" groups, such as doctors, for example, also have to be limited to a 48 h working week and 11 h daily rest period.

The EU driving time rules (for heavy traffic) restrict the total driving time to 9 h each day (on 2 days each week 1 h of extra driving time is allowed) (Directive 2002/654/EEC). A break (of at least 15 min) must be taken after not more than 4.5 h of driving. During the working shift a total of at least 45 min of break must be taken. Note that, apart from driving time, in practice it is possible to work a further 4 h on things like paperwork or loading, for example, without encroaching on the 11 h daily rest period. The weekly rest period has to be taken after not more than 6 days' driving and must last for at least 45 h. A reduction to 36 h can be made at the driver's domicile or to 24 h elsewhere—provided that compensation in time is given within 3 weeks. The total working time over a 2-week period is restricted to 90 h. The daily rest period has to be at least 11 h, but may be reduced to 9 h three times a week if compensation in time is given no later than the week following.

The United States does not have any general regulation on working hours—other than that 40 h/week is considered full-time work. Instead, some particularly sensitive areas have a specific Hours of Work regulation. One such area is road transport with heavy vehicles. Thus, the Federal Motor Carrier Safety Administration has hours of service regulations for commercial motor vehicle drivers carrying goods (slightly different for passenger drivers). This includes a maximum driving time of 11 h (after 10 h off duty). A driver may not drive beyond the 14th consecutive hour after coming on duty, following 10 consecutive hours off duty. The weekly limit is 60–70 h on duty, and a driver may restart a period of 7 or 8 consecutive days after taking 34 h off duty. Drivers using a sleeper berth must spend at least 8 consecutive h in the sleeper berth, plus a separate 2 consecutive h in berth, off duty.

It should be stressed that very little emphasis is laid on work during the night hours (the "diurnal low"), which may be the most important factor to regulate from a safety point of view.

8.5 DIURNAL RHYTHM AND SLEEP: THE MAIN COMPONENTS

The health effects of working hours that disrupt the rhythm deal with such things as sleep, wakefulness, accident risk, cardiovascular disease, and stomach and intestinal disorders. They have a close link to diurnal rhythm and sleep. In this section, we therefore introduce these concepts.

8.5.1 DIURNAL RHYTHM

Almost all physiological and psychological functions demonstrate a rhythmical behaviour over a 24-h period, that is, the level fluctuates between low and high values with a periodicity of 24 h [Czeisler and Dijk 2001]. The parameters employed to describe the rhythm are the same as those for a sine function; that is, the amplitude (the difference between the highest and lowest value), phase (the point in time of maximum value), and length of the period (the time between two peaks, that is, 24 h under normal circumstances).

Despite the fact that the basic diurnal pattern is approximately the same for most functions, there are major differences in timing. The hormone cortisol (see Fact Box 8.1), for example, reaches its highest level early in the morning; body temperature peaks in late afternoon and the pineal gland hormone melatonin peaks during the night (Figure 8.2). Generally speaking, however, most rhythms reach their maximum values during the day. It might be of interest to know that the bladder fills up four times as fast during the day as it does at night—which can easily disturb morning sleep. In addition, the stomach's ability to digest food is at its lowest late at night and at its highest in the afternoon. Late in the night is therefore the wrong time for food, that is, difficult to digest (fat, proteins). Susceptibility to medicines varies with the time of day, and late at night is, moreover, the point in time when most people die or are born (spontaneously).

FACT BOX 8.1

Cortisol is a hormone produced in the adrenal cortex on a signal from the pituitary gland in the brain. The aim is, among other things, to release more energy in detrimental situations, and cortisol is therefore often used as an indicator of stress. The hormone has a very marked diurnal rhythm with a peak at around 5 am. Melatonin is another hormone produced in the pineal gland. It has a marked diurnal rhythm that reaches its maximum between 4 am and 5 am. The hormone conveys signals from the biological clock out into the body and lowers the metabolism. This last helps melatonin to facilitate sleep. Melatonin also affects the setting of the biological clock, and is sometimes used to treat individuals who have maladjusted biological clocks.

The adaptation of the diurnal rhythm to night or morning work is a difficult one. Not even constant night workers can completely adapt their physiology to night work. What they need, of course, is help from daylight, and that occurs at the "wrong" time (i.e., during the day) in relation to night work. When people come out into the light from a night shift (around 6–7 am), their biological clock will stop the delay in the rhythm which light exposure during the night shift would have produced. The result is that melatonin, cortisol, body temperature, and other diurnal rhythm-regulating physiological variables are retained in their settings for day work. Often, however, we find a slight permanent delay in the diurnal rhythm of shift workers. An exception is night workers on oil rigs in the North Sea, where workers are not exposed to light to any great extent (they work indoors). There is actually an adaptation in that case, as the diurnal rhythm is adjusted for the indoor lighting, which of course follows the work-rest cycle.

If, on the other hand, we travel to another time zone, we are helped by the daylight of the new time zone, and the changeover then occurs relatively quickly. Over 4–5 days we will have adapted, for example, to New York time (westwards 6 h later than in the central part of Europe). It takes considerably longer to adjust to a flight eastwards, that is to time zones that lie ahead of us. This means that we have to shorten our day; that is, we have to go to bed earlier. This is less physiological—we are better constructed for flying west; that is, delaying the cycle.

A particular problem is that clock structures are also to be found in other organs—for example, in the liver, kidney, lungs, and heart. People need signals from the SCN, but the peripheral clocks are nevertheless relatively automatic. If we reset our diurnal rhythm, the biological clocks of the various organs will adapt themselves and "their" target organs at speeds that differ from each other. The long-term effect of this desynchronization is unclear, but it is assumed to be negative.

Being a morning person means you have an early maximum in your diurnal rhythm—both for body temperature and wakefulness. This means that sleep during the daytime is more difficult for the morning person. There is also, of course, an early trough—sleepiness comes earlier during the night shift. The circumstances for the evening person are the opposite. Above all, early mornings are difficult.

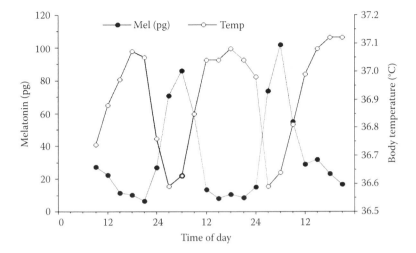

FIGURE 8.2 Diurnal rhythm for body temperature (temp) and melatonin (mel) over a period of 60 h of wakefulness; pg = picogram.

The anatomical structure that regulates the diurnal rhythm is to be found in the front lower part of the hypothalamus—just above the optic chiasma, and is called the "suprachiasmatic nucleus" (SCN). This biological clock consists of ~10,000 cells, each of which are controlled by their own genes. Some of these are called *Per, Tim, and Clock*. The genes are active ("express" themselves) during the daytime. This results in the production of proteins in the periphery of the cells. The proteins are carried back to the cell core and there inhibit any further "expression" by the genes. This cycle of activation and inhibition is what comprises the clock mechanism—the time is approximately a 24-h period. When the activity has reached zero, the next cycle begins. The whole group of cells together help each other to maintain a stable rhythm.

The cells of the biological clock also receive direct light information from the retina, which is used to "adjust" the setting. Without light information, the clock starts to "run slow"—it gets slower and out of time with the alternation of day/night. Light has its greatest effect on an individual who is awake immediately before or after the trough in the diurnal rhythm (of body temperature), around 5 am. Light before this point (and some hours beforehand) leads to a good hour's delay in the rhythm. The clock interprets light at this time as an extension of the day—which is why it delays its setting. Light after this time (and 6 h beyond that) leads to a corresponding advancement—light is regarded as an early sunrise. In this way, we adapt to different time zones and to winter- and summer time. The strength of the light also has significance. Usually, indoor lighting is rarely sufficient—if one does not spend a large part of the day in darkness and only has light during the critical hours. The clock can also be affected by activity (via serotonin and neuropeptide Y) or by food intake (via the appetite hormones leptin and ghrelin). But it is light that is the most important timer. The literature often uses the term "Zeitgeber" or "Synchronizer."

8.6   MEASURING THE DIURNAL RHYTHM

Measuring the diurnal rhythm for various bodily functions really involves simply deciding which function one is interested in, and then measuring that one a sufficient number of times to be able to form a view about the appearance of the rhythm. Among the favourite variables for measurement is body temperature, which is measured using a temperature sensor inserted up to 10 cm into the rectum and connected to a portable physiological monitor. The reason that body temperature is interesting is that it reflects the body's metabolism, and that physiological and psychological capacity are improved by high metabolism and impaired by low metabolism. Sleep has the opposite relationship—it is improved by low metabolism and impaired by high metabolism.

Favourite variable number two is probably melatonin (see Fact Box 8.1). This is governed directly from the SCN, supplies maximum levels at around 4 am, and is very low in the daytime. This lowers the metabolism and is affected (suppressed) by exposure to light. It is often used as a measure of the setting of the biological clock. By measuring the melatonin in saliva once an hour under constantly faint light conditions between 7 pm and 12 pm, one can estimate the setting of the SCN. Some time after 7 pm a steep rise occurs in melatonin secretion—"dim light melatonin onset" which is used to estimate the point in time at which the peak of the rhythm occurs. The techniques for analysing melatonin in urine and blood have existed for a long time. Today methods of analysing melatonin in saliva have become very popular among researchers into diurnal rhythms, because of the ease of acquiring samples.

Favourite variable number three is cortisol (see Fact Box 8.1). The function of cortisol has been discussed in Chapter 7 and its relation to stress is well known. It has a very stable diurnal rhythm, which also reflects the setting of the SCN. The level is lowest before, and, in particular, during the first hours of sleep, when active inhibition prevents secretion. Then the level rises towards point in time around awakening. For a long time there have been methods for analysing cortisol in blood and urine; in recent years, analysis of cortisol in saliva has become very popular among researchers in stress and diurnal rhythms.

Measurements of diurnal rhythms in the work environment are often imprecise, as, for example, melatonin is affected by light at the work place, and body temperature is affected by physical activity. One solution that is sometimes successful is to measure light and activity simultaneously. This can be used to adjust the values acquired. Anyone interested in this can acquire an activity meter (actigraph—see Section 6.12.4 in Chapter 6) which measures activity levels and which can be complemented with a light meter (and other monitoring instruments). Today activity meters are used a great deal in the field research into diurnal rhythms, stress, and sleep.

In certain exclusive laboratory trials, temperature is measured continuously in the rectum, and blood samples are taken every 20 min via a catheter. This last requires almost heroic efforts with regard to timing in order to ensure that the flow is not blocked and to provide follow-up treatment (e.g., centrifuging) of the blood and freezing it. Studies of this kind take many days and therefore make great demands on the logistics of monitoring. Often the subject is isolated from all timers so that the dynamics of the diurnal rhythm can be studied. This means that the experimenters must not give signals about the time of day—not even subtle signals such as appearing unshaven at 5 am.

8.7 SLEEP

Sleep is an altered state of consciousness during which the perception of external stimuli is strongly reduced, and a conscious act is impossible. Physiologically it is a state of reduced metabolism and increased anabolism (physiological construction). Characteristic changes are lower body temperature, heart rate, and blood pressure, at the same time as the secretion of growth hormone and testosterone increases, as well as the activity of the immune system.

The physiological signs of sleep are registered using an electroencephalogram (EEG), electrooculogram (EOG), and electromyogram (EMG). The EEG depicts brain activity via electrodes attached to the scalp; the EOG describes eye activity via electrodes attached near the outer corners of the eyes; and the EMG describes activity in, for example, the muscles of the chin by means of electrodes attached to the skin.

During sleep the EEG shows slow, but large, wave movements, increasing in size with increased depth of sleep. The EOG shows no activity before falling asleep (with closed eyes) but shows slowly billowing waves (slow eye movements (SEM)) on falling asleep. It also shows rapid, jerky movements when the subject is dreaming (rapid eye movement (REM) sleep). EMG activity is high when the subject is awake and declines slowly the deeper the subject sleeps. REM sleep is characterized by very rapidly falling EMG activity.

Sleep is divided into five different stages (Figure 8.3). Stage 1 is a transitional phase from wakefulness to sleep and has no value for recuperation. Stage 2 is "base

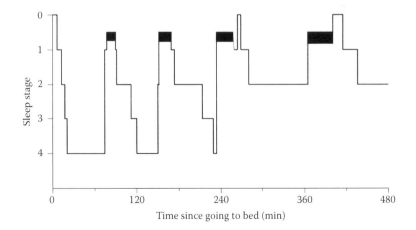

FIGURE 8.3 Sleep progresses during the night in a wave movement, from awake (stage 0), shallow sleep (stages 1 + 2), to deeper sleep (stages 3 + 4), to dream sleep (REM = blackened, to shallow sleep, etc.). A total of 4–6 cycles of this kind are gone through each night.

sleep" which takes up half of the sleep period. During this stage we recuperate but not to the maximum. Stages 3 and 4 are deep sleep with maximum recuperation. This is when we are also most difficult to wake up and are confused if we are woken. During REM sleep, which deviates considerably from other sleep, the brain and the rest of the body show raised activity, but strangely enough the postural muscles are relaxed (we cannot stand up or sit up). It is at this sleep stage that dreams are produced, even if dream images sometimes can occur in, for example, stage 2. We dream between four and six times every night. Deep sleep is always prioritized and is able to dominate the first half of sleep (see Figure 8.3). REM sleep is then withheld and is not released in earnest before the second half of sleep.

Disturbed sleep (insomnia) is characterized by a long period of falling asleep (more than 30 min), an increased number of awakenings (more than four), and/or relatively long periods of wakefulness during sleep (a total of at least 30 min). Often stages 3 and 4 are also reduced. The former often comprise the main criteria for sleep disturbance in clinical contexts. The diagnosis manuals for sleep disturbance also require that the problem should exist for at least a month so as not to be regarded as temporary. It is also necessary that the sleep disturbance should have consequences in the form of fatigue, irritability, or reduced functional capacity.

The function of sleep has not been fully explained. Sleep is a necessary precondition for activities while awake in the short term and for life itself in the long term. The minimum requirement of sleep seems to be around 7 h—in the long term. In the short term—from day to day—a sleep reduction from 8 to 6 h only has a marginal impact. After a reduction by 3 h one does, however, notice certain effects on the level of wakefulness and behaviour, and after the loss of a complete night's sleep the person affected shows a noticeably reduced capacity, comparable with the effects of sleeping pills. Three nights without sleep results in an almost total inability to carry out normal tasks requiring attentiveness, mental activity, or decision-making. As regards mortality, in the very long term this is somewhat higher among individuals who regularly sleep <4 h or more than 11 h a day.

Lack of sleep is compensated for primarily by more intensive (deep) sleep. For every hour we are awake we have to pay back ~3 min of deep sleep (stage 4). If, for example, we lose a night's sleep, we retrieve what we have missed largely by increasing the depth of our sleep—the length of sleep plays a minor role. A lost night's sleep can be regained the next night without our sleep being any longer! Increased depth is enough. Dream sleep seems to be considerably less important in the short term, and the lack of dream sleep does not seem to be made up. Long periods of suppressed dream sleep do, however, increase the pressure on recuperation, and there are signs indicating that dream sleep is important for the memory in the longer term. Research findings into the different stages of sleep are still not complete, but it seems as if deep sleep (stages 3 + 4—often called "slow wave sleep,") is necessary for the ability to keep oneself awake and for the immune defense system. REM sleep, on the other hand, seems to be associated with metabolism and temperature regulation. Both kinds of sleep seem to be important for memory consolidation.

Recently, results have been presented from different quarters showing that experimentally curtailed sleep leads to raised levels of the stress hormone cortisol and of blood fats (triglycerides) and to an impaired ability on the part of insulin to carry blood sugar to the cells. One working week with sleep cut by half leads to blood sugar levels above those necessary for diagnosing type 2 diabetes. The same observations have been made in patients suffering from insomnia or sleep apnoea (suspension of breathing during sleep).

8.8   SHIFT WORK AND SLEEP

One might discuss which physiological and health effects are most important in shift work. The clearest and most obvious effects are, however, those which affect sleep and wakefulness [Åkerstedt 2003]. It also seems to be those factors that determine whether one can cope with unusual working hours or not. The latter try to leave shiftwork. This process of elimination indicates that individuals with considerable experience of night work have fewer problems with sleep and wakefulness than would a non-selected group. Disturbed sleep and wakefulness in connection with shift work are included in a diagnosis group called "Shift Work Sleep Disorder" in the international diagnosis manual for sleep disturbances [AASM 2005]. This identifies a large group of people who suffer from problems in sleeping or from "nonrestorative sleep" on most days when working night shift—but otherwise sleep normally. This group has an increased probability of not coping with shift work, and has a greater risk of, for example, accidents, absence due to illness and cardiovascular disease.

The effects of shift work on sleep have been fairly well researched. Field surveys of three-shift workers and similar groups have shown that sleep disturbances are very common. At least three-quarters of the shift workers were affected by sleep disturbance. EEG studies of sleep among workers with rotating shifts show relatively consistent results, irrespective of whether these are carried out in the laboratory or in the home. Sleep during the daytime and sleep before an early-morning shift is 1–4 h shorter than night sleep in connection with daytime work. The reduction primarily affects stage 2 and REM (dream sleep). Stages 3 and 4 (deep sleep) seem not to be affected. In addition, sleep onset latency is abbreviated in day sleep after a morning shift, and REM sleep often sets in earlier, though not always.

Often the sleep pattern before the morning shift does not vary between individuals, that is, most people go to bed and get up largely at the same time. In relation to evening or night work, the variation between individuals is considerable. Morning shifts are also characterized by the sleep quality being disturbed by something that is very similar to stress. The more one is worried about the difficulty of waking up, the less deep sleep (stages 3 + 4) is produced. The subjective effects that mark morning shift sleep have to do with non-spontaneous waking (almost everyone uses an alarm clock or some other method) and a feeling of not being thoroughly rested. Approximately one-third of night workers supplement their main sleep with a nap, which is often a direct function of the abbreviation of their main sleep.

One would imagine that working for several nights in a row would mean that sleep improves day by day. This is not normally the case, however. Sleep curtailment remains—even for individuals with permanent night work. There are sometimes tendencies to adaptation, particularly in relation to permanent night work, but the effect is very limited. The reason is, as discussed earlier, that the adjustment of the biological clock to night work is counteracted by exposure to daylight on the morning after the night shift.

8.9 CAUSES OF SLEEP DISTURBANCE

Several earlier studies have pointed to the fact that noise levels are higher during the day to explain why daytime sleep is shorter after a night shift. On the other hand, sleep after a night shift is shorter even in the quietest laboratory environment. So, noise does not seem to be the main cause of disturbed daytime sleep.

As has been mentioned previously, a stronger influence is exercised by the biological diurnal rhythm. If we postpone going to bed to a time other than the optimal (about 11 pm–7 am), our sleep will be negatively impacted (Figure 8.4). The later one goes to bed (12 pm, 2 am, 4 am, etc.), the shorter the sleep will be—until bedtime the next day around 4–6 pm, when the length of sleep will begin to increase. The reason is that the switching point of the biological clock (around 5 am) is followed by increasing metabolism. In this way, the sleep process is disrupted and one wakes up early. If one stays awake until the next evening, the diurnal rhythm will have turned (round about 5 pm) and the disrupting effect begins to subside.

It has also been shown that it is not merely the time of waking that is regulated by the diurnal rhythm, but also the time of going to bed (if social influences are limited). The maximum readiness for sleep coincides approximately with the diurnal trough in body temperature, but usually the first signs of approaching sleep appear a couple of hours earlier. It is interesting to note that the "going to bed signal" is preceded by a short phase of great wakefulness, when it is very difficult to sleep.

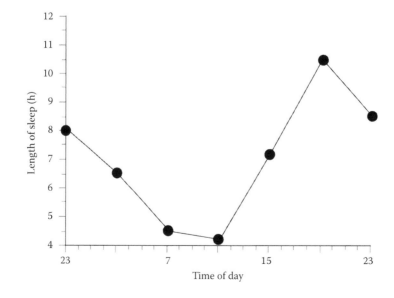

FIGURE 8.4 Length of sleep after going to bed at different times of the day. Note that at 11 pm on the left one has been awake for 16 h and at 11 pm on the right one has been awake for 40 h.

8.10 EFFECTS OF WAKEFULNESS, PERFORMANCE, AND SAFETY

8.10.1 EFFECTS OF WAKEFULNESS AND HOW TO MEASURE THEM

Most surveys show that three-shift workers and similar groups experience fatigue more often than day workers [Åkerstedt 2003]. Normally, fatigue is general during the night shift, scarcely noticeable during the afternoon shift and of short duration during the morning shift. Certain studies report cases where sleepiness has been so great that it has led to people falling asleep during the night shift.

The methods of measurement for sleepiness/wakefulness are largely a matter of subjective assessments. One of these is the so-called visual analog scale for sleepiness, which is usually presented as a 10-cm long line with "very alert" and "very drowsy" at its endpoints. Likert scales are also popular. They have a number of steps, usually 7 to 9, where each step is based on the description of the phenomenon of sleepiness. An example is the "Karolinska Sleepiness Scale" which has the following basis: 1 = extremely alert, 2 = very alert, 3 = alert, 4 = quite alert, 5 = neither alert nor drowsy, 6 = first signs of sleepiness, 7 = drowsy but no effort to stay awake, 8 drowsy some effort to stay awake, and 9 = very drowsy + major effort to stay awake + fighting sleep (see also Figure 8.5 in which this scale is used).

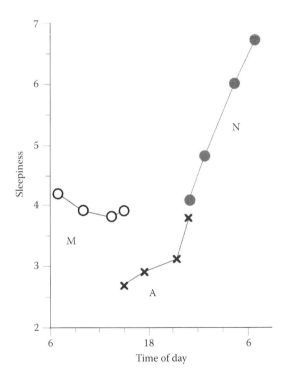

FIGURE 8.5 Self-assessed sleepiness during the morning (M), afternoon (E), and night (N) shift—with time between shifts eliminated so as to illustrate the diurnal pattern. Sleepiness assessed on the Karolinska Sleepiness Scale of 1–9, where higher values indicate great sleepiness.

Another popular model is the so-called multiple sleep latency test which measures readiness for sleep at several times during the daytime (after a normal night's sleep). The subject is provided with electrodes for the EEG and EOG (in the same way as for sleep), placed in a dark room and asked to lie down, close their eyes, and not fight sleep. The time from going to bed up to the first 20-second interval with sleep comprises the measure (in minutes). The trial is interrupted after the first sleep, or after 20 min if sleep has not occurred (the maximum value is then 20). Normal values lie between 10 and 20 min. The criterion for suspecting unhealthy fatigue is a time to falling asleep of <5 min. At the end of a night shift, values of around 2–3 min are often measured.

Other variants quite simply comprise the number of sleep events in EEG and EOG during work or under controlled situations during a break from work. Sleep events chiefly mean all activity within the frequency range of the EEG signal that one knows is related to sleepiness or sleep. This means, above all, the interval between 4 and 12 Hz (α and θ activity). The EOG then generally shows slow undulating eye movements or at least very slow blinks. The latter means that the eyelid stays closed for more than 0.15 s. Active wakefulness produces EEG frequencies of more than 12 Hz (β activity). A number of studies show that increases in SEM and α and θ activity in EEG are directly related to sleep and behavioural interruptions in the interaction with the surroundings (often called "lapses" in the literature) just as they are to the experience of sleepiness.

Figure 8.5 shows what self-assessed sleepiness looks like during a morning, afternoon, and night shift. The graphs have been produced for different days but collated to illustrate the clear pattern of the diurnal rhythm. Sleepiness is lowest during the afternoon shift, average during the morning shift, and very high towards the end of the night shift.

There are few physiological studies of sleepiness in connection with night-shift work. In a study carried out by Torsvall et al. [1989], however, EEG and EOG were registered at work. This was done with the help of a small recorder carried by the subject during a 24-h period including the morning, afternoon, or night shift. During working hours, a quarter of the participants showed sleep patterns in the EEG. This occurred in most cases during the second half of the night shift, and never in connection with any other shift. It is important to emphasize that the company did not sanction sleeping on the job; nor were they aware that this was taking place. Similar studies have been carried out on locomotive drivers, truck drivers, and pilots. All these groups show clear signs of sleeping at work (α and θ activity, SEM). In the introductory section, George showed a greater activity primarily in the α band, and, furthermore, slow blinks at about 3–4 am when he had been driving for 7–8 h.

Data on how many people suffer from sleepiness vary somewhat, depending on exactly what is meant by the concept. Most night workers do, however, report greater sleepiness during night work and morning work. More than half report at least one occasion of falling asleep at work, and 10–20% experience this in connection with every night shift.

The underlying causes of these effects on wakefulness are primarily the length of time awake, the length of the previous sleep, and the phase (point in time) in the diurnal rhythm. The shorter the previous sleep has been, the longer one has been awake and of course the closer to the trough (4 am–6 am) in the diurnal rhythm one finds oneself, the more severe the sleepiness becomes. The combination of these effects has been modeled mathematically, and today there are several such models to predict future sleepiness levels. These models generally also predict performance and safety, which are discussed below.

8.10.2   EFFECTS ON PERFORMANCE AND SAFETY

If sleepiness at the workplace is as extensive and as dramatic as has been suggested above, one might expect an impact on performance and consequently on productivity and safety. A classic in this field is a study carried out by Bjerner et al. [1955] showing that the incorrect readings taken at a gasworks over 20 years had a clearly marked peak during the night shift (Figure 8.6). A less prominent peak also occurred during the afternoon. Similar results have been demonstrated for performance among locomotive drivers and the response time for telephone operators. But generally speaking, few effects can be found on performance. The reason is presumably that a great deal of work is machine controlled, and that rather little of the production depends on individual performance.

A large number of studies have also been carried out using different types of performance testing. Most show clear effects of night work on, for example, reaction times and different types of what are called tracking test (following a movement with some kind of joystick or steering wheel). Laboratory studies also very consistently indicate clear negative effects on performance in a number of psychological tests. One method for summarizing the effects is to compare them with the effects of alcohol. Several studies have been carried out, and work late at night

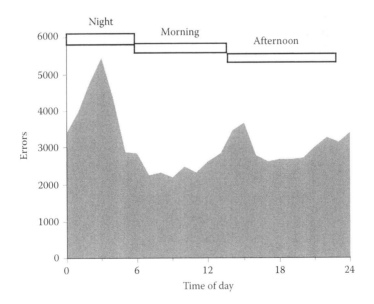

FIGURE 8.6   Effect of time-of-day/shift on incorrect readings at a Swedish gasworks.

seems to entail a drop in performance to the same extent as having a blood alcohol level of 0.8 per thousand.

One serious effect of fatigue is a reduction in safety. With a serious lack of sleep, the interplay between the individual and their surroundings is interrupted. If this coincides with a situation that requires action, an accident may occur. The transport sector is the area where the connection between shift work and accidents appears most clearly. The US National Transportation Safety Board has drawn the conclusion that fatigue is one of the most important causes of accidents within the transport area (15–30% of all accidents), and that fatigue is the most important individual factor in accidents involving heavy traffic on the roads. It is primarily single-vehicle accidents that multiply in night driving [Philip and Åkerstedt 2006]. But all kinds of road accidents, apart from accidents while passing, increase late at night.

An interesting analysis has been presented by the Association of Professional Sleep Societies' Committee on Catastrophes, Sleep, and Public Policy. The report draws attention to the fact that the core meltdown of the nuclear power station at Chernobyl in the then Soviet Union occurred at 1:35 am, and was caused by a fault resulting from human error (clearly related to the scheduling of work shifts). In the same way, the accident at the nuclear power station at Three Mile Island in the United States occurred between 4 am and 6 am, and was caused not merely by the fault in the valve which meant that there was a leakage of cooling water, but primarily by the fact that no one had noticed this occurring. Similar incidents which, however, were prevented at the last minute, occurred in 1985 at the David Beese reactor in Ohio and at the Rancho Seco reactor in California. Finally, the committee also observes that the catastrophe involving the NASA space shuttle *Challenger* originated in errors of judgment made early in the morning by people who had not had sufficient sleep (as a result of part-time night work) in the days leading up to the launch. It should be stressed that almost the entire official attention in connection with these accidents has been focused on their technical aspects.

Despite convincing research results, knowledge about fatigue as a cause of accidents has is seldom considered in the work on safety at workplaces, or in traffic safety work. The reason is, presumably, partially a question of tradition within the responsible organizations and partially the difficulty of measuring fatigue in real situations and to take appropriate action. The sleep latency test, by the way, is not suitable for field use by the side of the road since being stopped by police is likely to enhance alertness.

As regards monitoring methods for mental performance, at least, then presumably the simple serial reaction time test is the one most often used and also the most sensitive to a fatigue-related drop in performance. This means that anyone who is to be studied is given a series of signals (a lamp lights up) at intervals of 2–9 s. The task is immediately to press a corresponding button. The tracking test in various forms is also sensitive. This means that, using a joystick or the like, the subject has to follow a dot on a screen which moves unpredictably. Both these types of tests have been used in field studies. But, just as in the case of physiological tests, "masking" occurs, that is to say conditions such as the work situation, lighting, or noise may easily impact on the performance.

8.11   OTHER HEALTH EFFECTS

The longer-term effects on health primarily concern mortality, cardiovascular disease, disease of the stomach and intestines, cancer, depression, and disruptions in pregnancy.

8.11.1   MORTALITY

Mortality among shift workers has been investigated very little. In one of the few (careful) studies, Taylor and Pocock [1972] compared mortality in a group consisting of 8603 day and shift workers over a period of 13 years. The only difference that could be proved was that former shift workers have a higher mortality than the then current day or shift workers. The reason for the lack of studies is presumably the turnover among shift workers. It is difficult to gain a clear picture of the amount of shift work that has taken place, not least among day workers, as most studies of mortality are conducted afterwards, and registers in general lack sufficient information about previous working hours. Later studies have shown similar results.

8.11.2   DISORDERS OF THE STOMACH AND INTESTINES

Studies show that irregular working hours, and night-shift work in particular, are associated with disorders of the stomach and intestines, for example, loss of appetite, constipation, "gas," and heartburn. One of the most systematic studies (34,047 individuals) showed that stomach ulcers were 10 times as common in two-shift workers than day workers, and 12 times more common in three-shift workers [Angersbach et al. 1980]. Among former shift workers, stomach ulcers were ~20 times more common. The differences between shift workers and others began to appear ~5 years after hire.

The mechanism underlying the problems is presumably that the diurnal pattern for stomach enzymes and for the stomach's motor patterns does not correspond to the sleep/wakefulness pattern. The pattern of food intake is, furthermore, irregular. A high nocturnal intake of food seems to be linked to increased blood lipid levels. Eating at around the time of the trough in the diurnal rhythm seems to cause an altered metabolic reaction, among other things, a reduction in the ability of insulin to store blood sugar in the cells (more blood sugar remains in the blood). The energy metabolism is also different at different times of the day. On the other hand, there are today no data linking the stomach ulcer bacterium *Helicobacter pylori* with shift work.

8.11.3   CARDIOVASCULAR DISEASE

Cardiovascular disease is one of the more important areas for the study of the effects of working hours. The effects appear clear and many outline studies have been published. In a study of a paper manufacturer, the incidence of heart attacks between 1968 and 1983 was reported for a group consisting of 504 shift and day workers, and this was related to how great the exposure (number of years) had been for shift workers [Knutsson et al. 1986]. A few of the original participants were not traceable. The age-standardized relative risk for shift workers compared with day workers was 1.4 (i.e., 40% excess risk). It was not possible to determine the risk for workers with <2 years exposure (no one in that group had a heart attack). In the group of workers with 2–5 years, exposure of the risk was 1.5; for 6–10 years' exposure, it was 2.0; and for 16–28 years' exposure, it was 2.2 and 2.8, respectively. In the group with more than 21 years of shift work, the risk dropped to 0.4. This last is assumed to result from a selection effect—the most sensitive workers had left shift work.

The reason behind the excess risk among shift workers is not known, but may have to do with the raised blood lipid levels which are often to be found among shift workers. This may be the result, for example, of too great a food intake at night, poor diet, or sleep disturbance, and so on.

8.11.4 OTHER ILLNESSES

Several studies have shown a certain excess risk of low birth weight among the children of shift-working mothers, and there are signs of a higher proportion of spontaneous abortions. Shift work also seems to increase the risk of cancer, particularly breast cancer in women and prostate cancer in men. The reason is unclear, but it has been proposed that exposure to light at night leads to a suppression of melatonin secretion during the night shift, and that the anticarcinogenic effects of the melatonin are therefore reduced. Diabetes also appears to have a link to shift work, even if the scientific support is still as yet limited.

8.12   INDIVIDUAL DIFFERENCES

There are quite clearly major individual differences in one's ability to tolerate night shift work. Often one finds that the majority of workers have very few problems, while the group with major problems scarcely exceeds 10%. There is also a group that is at least as big that has no negative reactions at all from shift work. Many attempts have been made to identify special risk groups who might potentially suffer from unusual working hours. No clear conclusions can be drawn, however. There are scarcely any gender differences. Increased age may possibly mean greater difficulties, primarily with daytime sleep (the sleep mechanism is somewhat weaker in older individuals, and the diurnal rhythm's waking effects take over). Night sleep is, of course, also disturbed by increased age, but relatively modestly—waking more often and waking up earlier, again resulting from a weaker sleep mechanism. This leads to a greater difficulty in coping with shift work with increased age; 45 years of age seems to be a turning point. One interesting observation that has been made in several studies is that older people are less sensitive than younger ones to sleepiness when taking the odd night shift.

Morning people have problems in relation to night shifts, but on the other hand have advantages on morning shifts. In all likelihood, it is also true that individuals with a tendency to sleep disturbance have greater problems with night work, as the sleep that follows encounters resistance from their biological clock. This has, however, not been studied.

One observation which may be relevant is that those individuals who are most physiologically adapted to night work (high melatonin levels during the day and low during the night) have fewer problems. This presumably reflects the greater focusing on night work on the part of the individual and possibly a more successful approach to dealing with daytime sleep. This last may mean that they black out the bedroom more efficiently, that they avoid exposure to light on the way home from their night shift (with the help of dark glasses) and that they also have an evening orientation in their lives on their days off. These are speculations, however, which still have to be tested.

One study recently investigated what characterizes shift workers with a very negative, or respectively very positive, attitude to their working hours. A number of factors were tested, and in brief the strongest factor was much greater fatigue in relation to night work in combination with a lack of recuperation during sleep [Axelsson et al. 2004]. The sleep was, however, not curtailed. Ongoing research is being pursued in order to determine what factors may be involved in fatigue.

8.13 SPECIAL FACTORS CONCERNING WORKING HOURS

There are a number of aspects of disruption in the diurnal rhythm that have to be taken into account when we talk about the effects of unusual working hours. One of these is the time between shifts. Evidently, extremely rapidly rotating shifts (with 8 h between shifts) do not provide sufficient time for recuperation. A somewhat slower rotation—2–3 similar shifts in a row—seems to be a good compromise. Longer sequences of night work increase fatigue (physiological adaptation does not occur).

Another factor is the direction of the rotation. As postponement seems to be an inbuilt tendency in the circadian rhythm when synchronizers are not in evidence, it is natural to allow the work shifts to alternate from morning to afternoon and then night. Going from an afternoon shift to a morning shift on the other hand means that the time off between shifts is shortened to 8 h, which results in greater fatigue. Backward rotation is, however, often the wish of the employee to achieve a longer continuous time off.

Long shifts are often physiologically unproblematic if they do not exceed 12 h, as long as there are breaks and the workload is not too high. This also applies to the transport field—a long period of wakefulness, reduced sleep, or night work is considerably more important causes of fatigue. As mentioned earlier, extended work shifts are a marked trend in Europe—to provide long continuous periods of time off.

The number of days in a row which can be worked without physiological effects is unknown, but there are studies indicating that the risk of errors rises markedly among locomotive drivers after 7 successive days of work. Some studies have shown similar effects for industrial work.

8.14 RISK-REDUCING MEASURES

Working at night is in itself not physiological. It can logically never be equivalent to day work as regards strain, but as has been discussed above, there are factors that can improve the situation.

Presumably the most important countermeasure is reduced work hours. Fatigue needs to be counteracted with sleep and a normalization of the diurnal rhythm. Then an extra day is needed to adjust from night work.

It may also be that a shift longer than 8 h may help (paradoxically enough), as fewer days are in this way allocated to night work; the worker has more days with natural (night) sleep and avoids early morning shifts which also constitute a problem of difficult and non-physiological waking times. But the work load also has to be adapted so that the worker can cope with long shifts, and there has to be the opportunity for breaks when the nature of the work requires it.

The days off work which form part of the shift compensation, or which are acquired through taking longer shifts, should of course be used for recuperation immediately once the load has ceased. That is where they are needed most. To moonlight during leisure time is inappropriate because it often prevents rest and recuperation.

One could discuss whether permanent night work might be an alternative to that system of alternation between shifts which now dominates. It is somewhat easier to deal with night work if it is made permanent. The disadvantage of permanent night work is, however, that one risks ending up being excluded from in-service training and other things that happen during the daytime. Employers should therefore ensure that day work is also undertaken.

In night shift work it is otherwise a good idea to have an early end to the shift. This makes for longer sleep. Ending the night shift at 5 am provides an almost normal length of sleep, in that one has time to get in around 6 h of sleep before the diurnal rhythm's wake-up effect sets in during the early afternoon (see Figure 8.4). The disadvantage is that the morning shift is affected by a reduction in sleep because of having to get up so early.

Individuals must themselves ensure that they optimize their sleep environment. This means a total blackout of the bedroom, low temperature (14–16°—air conditioning in summer) and a silent environment (earplugs are recommended). One should also ensure that incomplete daytime sleep is complemented by taking naps and restricting any moonlighting. A short (20 min) nap can easily replace 2 h of lost main sleep. The reason is that main sleep is very ineffective towards the end, while the nap is more effective. Shift workers should also be aware of their diet so as to counteract raised blood lipid levels.

The greater difficulty of daytime sleep with advancing age (as mentioned earlier) may often make it impossible to continue with night work. For this group and for others with recuperation problems in daytime sleep, there is today no alternative other than transferring to some form of day work (or stopping work completely). It is, moreover, wise to try to find an opportunity for alternation between timetables with and without night work, so as to allow employees who want to work at night to do so when they feel they are able and need to.

We should also be aware that treatment with pharmaceuticals to reduce fatigue has begun to be used in the United States. It is, however, unclear as to whether this is suitable in the long term. Nothing is known about the consequences, and it is doubtful as to whether fatigue, as a natural reaction to too little sleep, should be treated at all.

8.15 SUMMARY

George's sleepiness is, as we have seen, a safety risk when he is driving his truck. Night work and morning work cause a conflict with the normal day-oriented physiology. The result is problems in sleeping during the daytime and being awake at night time. In the long term this has consequences for safety and health. If someone is awake for more than 16 h or asleep for <6 h, there is a clearly increased safety risk. Shift work also has certain negative effects on long-term health. The function of sleep is to restore the brain's functional level after the reduction occurring during the waking hours of the day. Sleep is also a precondition for the rest of the body's physiology to be able to recuperate and function normally. There is really no way of eliminating these problems. On the other hand, we can alleviate the effects by sleeping strategically (a nap is extremely efficient), optimizing our sleep environment, and ensuring that we have regular recuperation. It is also important to be aware of the greater risk of accidents in connection with night work, and of the need for a healthy diet. Legislation has taken note of some of the problems by limiting the length of driving time, but on the other hand in practice it ignores (strangely enough) the effects of working at night.

REFERENCES

AASM ICSD. 2005. *International Classification of Sleep Disorders, Revised: Diagnostic and Coding Manual*. Chicago, IL: American Academy of Sleep Medicine.

Angersbach, D., P. Knauth, H. Loskant, MJ. Karvonen, K. Undeutsch, and J. Rutenfranz. 1980. A retrospective cohort study comparing complaints and disease in day and shift workers. *Int Arch Occup Environ Health* 45:127–140.

Åkerstedt, T. 2003. Shift work and disturbed sleep/wakefulness. *Occup Med* 53:89–94.

Axelsson, J., T. Åkerstedt, G. Kecklund, and A. Lowden. 2004. Tolerance to shift work—How does it relate to sleep and wakefulness? *Int Arch Occup Environ Health* 77:121–129.

Bjerner, B., Å. Holm, and Å. Swensson. 1955. Diurnal variation of mental performance. A study of three-shift workers. *Br J Ind Med* 12:103–110.

Czeisler, CA. and D-J. Dijk. 2001. Human circadian physiology and sleep–wake regulation. In: Takahashi, FW. and RY. Moore. (Eds), *Handbook of Behavioral Neurobiology*. New York: Kluwer Academic/Plenum Publishers, pp. 531–569.

Directive 2002/15/EC of the European Parliament and of the Council of 11 March 2002 *on the organisation of the working time of persons performing mobile road transport activities*. http://eur-lex.europa.eu//en/index.htm

Directive 2003/88/EC of the European Parliament and of the Council of 4 November 2003 *concerning certain aspects of the organisation of working time*. http://eur-lex.europa.eu//en/index.htm

Knutsson, A., T. Åkerstedt, B. G. Jonsson, and K. Orth-Gomér. 1986. Increased risk of ischemic heart disease in shift workers. *The Lancet* 12;2:89–92.

Philip, P. and T. Åkerstedt. 2006. Transport and industrial safety, how are they affected by sleepiness and sleep restriction? *Sleep Med Rev* 10:347–356.

Taylor, PJ. and SJ. Pocock. 1972. Mortality of shift and day workers 1956–68. *Br J Ind Med* 29:201–207.

Torsvall, L., T. Åkerstedt, K. Gillander, and A. Knutsson. 1989. Sleep on the night shift: 24-Hour EEG monitoring of spontaneous sleep/wake behavior. *Psychophysiology* 26:352–358.

FURTHER READING

Boggild, H. 2009. Settling the question—The next review on shift work and heart disease in 2019. *Scand J Work Environ Health* 35:157–161.

Sallinen, M. and G. Kecklund. 2010. Shift work, sleep and sleepiness—Differences between shift schedules and systems. *Scand J Work Environ Health* 36:121–133.

9 Work in Heat and Cold

Désirée Gavhed

Photo: Allan Toomingas

CONTENTS

Bert, who is 57, has worked full time for many years for a building firm in Canada. He works all year round, mainly outdoors. The firm primarily constructs office buildings and private houses. Bert's tasks vary. Sometimes the work is very heavy when there are building materials to be carried. At other times, it is fairly sedentary, when he supervises machine work, for example, when a hoist is moving materials.

Bert generally enjoys his work, but is starting to think that it is hard going with the heavy operations in very hot or very cold weather. He is not as physically active in his leisure time as he was 20 years ago, when he often went skiing during the cold season and went swimming a lot in summer. In winter, Bert feels a little stiff, and has pain in his knees and in two fingers which got frostbitten one particularly cold winter. During the days of summer he gets very tired after work and can even feel a little unwell when he is at his worst. In recent health checks, he was found to have high blood pressure, and he now has to take antihypertensive drugs on a regular basis.

9.1 FOCUS AND DELIMITATION

This chapter deals with how working in the heat and cold affects the body and productive capacity. The demands made on the body's physiological adaptability increase when it is subjected to external stress. In extreme conditions, such as doing heavy work in the heat, the body does not always manage to compensate for the considerable heat load, and the risk of injury and ill health is considerable. This chapter answers questions such as:

- Can Bert's bad knee have anything to do with the climate at his workplace?
- Why is Bert's pulse rate so high when he works in the sun?
- How can Bert protect himself against heat and cold?
- Are there any opportunities for anticipating what conditions people can work under in different climates, and for how long?
- Does Bert's age have any significance for his work ability in heat and cold?

9.2 PREVALENCE OF WORK AND DISORDERS IN HEAT AND COLD

Many people have problems with the climate at their workplaces. The problems are varied in nature and depend on whether it is a question of outdoor work, an office workplace, or work in industry. Problems with the climate at workplaces seem not to have been dealt with to any extent over the last 20 years, and there is little to indicate that exposure to climate will change to a great extent in the coming decade. The concept of *thermal climate* has to do with heat and cold, unlike the psychosocial climate. *Exposure to climate* is the dose of heat or cold, humidity, and wind speed to which people are subjected.

A large number of workers are exposed to cold and/or heat at their workplaces, and many of them experience cold and/or heat strain [Eurofound 2007]. The greatest proportion of disorders is reported from work that is mostly performed outdoors, manufacturing work, and transport work. About a quarter of the workers in Europe are exposed to low or high temperatures for a quarter or more of their workday. In all, 25% of the workers are exposed to "high temperatures which make them perspire even when not working," while 22% are exposed to "low temperatures whether indoors or outdoors." Male workers are more exposed to challenging thermal environments than female workers are, according to EU Working Conditions statistics (heat: 31% for males vs. 18% for females, low temperature: 29% for males vs. 13% for females). This is because a large proportion of men work in businesses involving either high or low temperatures such as building and installation, forestry and agriculture, and manufacturing industry. In the European Union (EU) member states, the reported percentage of workers experiencing thermal strain ranges from 14% to 45% [Eurofound 2007]. In order to reduce the number of disorders, thorough knowledge and practical measures are required.

9.3   WHAT CHARACTERIZES WORK IN HEAT AND COLD?

Exposure to extreme climates varies in both time and intensity in different businesses. In work in a large-scale kitchen, staff only go into the refrigeration and freezer rooms on a few occasions each day to fetch or leave food and stay there for a very short span. Building workers, as in our first example, can be exposed to extreme cold or sunshine and heat throughout their working day. Frequently, climate problems are related directly to the business and/or the outdoor climate. Examples are workplaces with heat-generating processes, for example, manufacturing pulp or casting metals.

The most common causes of climate problems in outdoor work are solar radiation in hot environments in combination with heavy work as well as wind and cold in cold climatic zones. In parts of the world where the climate varies a great deal between different regions and seasons, work has to be adapted to the specific situation. Even in a normal indoor climate (e.g., an office) climate-related problems often occur.

It is important that assessments of the risks of disorders are based on the real exposure to climate, that is to say how long and with what intensity the employee is exposed to heat or cold, and not merely on measurements of the climate at the workplace.

The following environmental and individual factors are of significance for the individual's well-being.

Climate	Individual
• Air temperature	• Heat production (which is dependent on work load)
• Radiant temperature	• Thermal properties of clothing
• Wind speed	
• Humidity	

9.4   HEAT PRODUCTION, HEAT LOSS, AND TEMPERATURE REGULATION

In the digestion of food, carbohydrates, fats, and proteins are metabolized to form energy-containing substances used for life-sustaining processes, work, growth, and recuperation (among others, ATP—see Chapter 2, Section 2.4). At rest, this energy metabolism is called *basal metabolism or Basal Metabolic Rate, BMR* (see Fact Box 9.1).

FACT BOX 9.1

OXYGEN CONSUMPTION, ENERGY METABOLISM

At rest, ~3.5 mL of oxygen is used per minute and kilogram of body weight, which for a person weighing 80 kg amount to 0.28 L of oxygen/min. The energy metabolism is ~6 kJ/min, which means that the heat output developing in the body is ~100 W/m^2 or ~55 W/m^2 of body surface at rest. The body surface of a "standard" person is ~1.7 m^2 (man 1.8 m^2 and woman 1.6 m^2).

At work, the stored chemical energy is converted into mechanical work and heat. The proportion of the energy not directly used for work (the efficiency) in dynamic muscle work with major muscle groups is at most 30%, but is negligibly small when small muscle groups are recruited (e.g., arm work). For example, when Bert is hammering nails, his efficiency is 15%, and 85% of the energy is therefore converted into heat. For many work operations, all the converted energy becomes heat (see Section 9.10.2).

The temperature in the body core (the central parts of the body containing the internal organs) is at rest normally around 37°C and usually varies ~0.5°C during the day. This is important because all biochemical processes in the internal organs and the brain function optimally at around 37°C. Other organs are also dependent on temperature; for example, the muscles function best when they are warm.

Body temperature, like many other physiological functions, has a diurnal variation, which can influence the experience of the thermal climate and its effects (see also Chapter 8, Section 8.5.1). The body heat content (and thereby temperature) is determined by the balance between heat production from the basal metabolic rate, heat production in the muscles during physical work, and the heat emitted by the skin surface. In a normal room climate there is a temperature gradient (a gradually changing temperature difference) from the body core to the body surface and from the body core to legs and arms, which are normally some degrees lower (Figure 9.1). The skin temperature therefore varies across the whole body, but is usually 33°C in the trunk at, for example, room temperature and wearing indoor clothes.

In muscle work, the core body temperature and the muscle temperature rise with increasing physical load (which should not be confused with a fever).

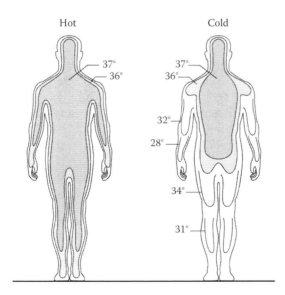

FIGURE 9.1 Temperature distribution in the body in hot and cold environments. Illustration: Niklas Hofvander.

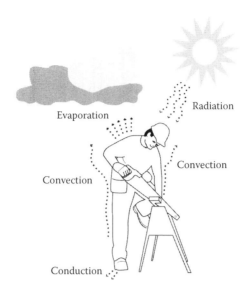

FIGURE 9.2 Heat exchange between the body and its surroundings. Illustration: Niklas Hofvander.

The heat from the body surface is emitted through convection, radiation, conduction, and evaporation and by respiration through convection and evaporation (see also Figure 9.2).

- *Convection (movements in air or fluid)*: the surrounding air or water is heated by contact with the skin. Warm air is lighter than cold and therefore flows upwards along the body and away from the body. Wind or movements of the body leads to increased convection; that is, more heat is emitted from the body surface.
- *Radiation*: heat is emitted through electromagnetic waves directly from the skin to colder surfaces in the surroundings (or vice versa through solar radiation for other heat radiation, e.g., from a furnace).
- *Conduction*: heat is transferred directly in the contact surface between the skin and the material.
- *Evaporation*: heat is emitted from the body when it accompanies the fluid evaporating on the surface of the skin and mucous membranes and away from the body. With high humidity in the surroundings or if tight clothes are worn, sweat and moisture evaporates poorly, and the cooling effect is reduced.

In water, which has heat conductivity 25 times higher than air, large quantities of heat are quickly transferred through conduction, for example, when diving. Heat transfer can also occur in the opposite direction, from the surroundings to the body surface, when the body is exposed to heat radiation or warm air. The temperature of the body surface may then approach, or in extreme cases exceed, that of the core.

The body strives for a balance (homoeostasis) between heat production and heat transfer (heat balance) to maintain an optimum environment for all the biochemical processes necessary to life. Temperature regulation involves thermoreceptors that are constantly measuring the temperature and which send a signal to the control centre in the hypothalamus in the central nervous system (CNS) (Fact Box 9.2). Certain receptors are stimulated by cold, others by heat. Deviations from normal body temperature activate regulators with the aid of hormones and the nervous system. Nerve signals are transmitted from the hypothalamus to those organs involved in temperature regulation (blood vessels, heart, sweat glands, adrenal glands, thymus, and skeletal muscles).

FACT BOX 9.2

The heat and cold receptors in the skin are neurons, which in their receptor portions have ion channels which open under temperature stimulation. It is perhaps noteworthy that the ion canals of the cold receptors can also be opened by chemical substances, for example, menthol, which explains why menthol produces a feeling of cold.

The most important physiological mechanisms of temperature regulation are:

- Blood vessel constriction/dilation
- Sweating
- Shivering
- Changes in metabolism

For more detailed studies of temperature regulation, see textbooks in physiology (listed in Further Reading).

What is generally meant by cold load or heat load is a situation where the body's temperature regulation is activated; see the subsequent sections on heat and cold. This is a result of the physical work load (which makes for increased body temperature) and how warm one's clothes are (which prevent heat from being emitted from the body) when the ambient climate puts strain on the body. As previously mentioned, other climatic factors also affect climate load on the individual. In this way, "heat" and "cold" are not precise temperatures. In the context of measurements and ratings, ambient temperatures below +10°C are termed as "cold," 10–30°C are called "indoor climate," and above 30°C are called "heat" (see Section 9.10).

9.5   COLD

9.5.1   PHYSIOLOGICAL RESPONSES TO COLD AND THE EFFECTS ON WORK ABILITY AND HEALTH

9.5.1.1   Rest

When Bert is supervising a building project in the cold, his body is protected by the circulation being redistributed from the peripheral body parts (hands, feet, ears, and nose) to the central parts of the body and the brain (see also Figure 9.1). The hands

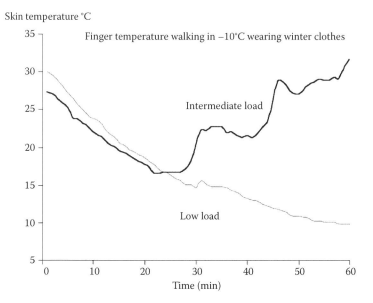

Skin temperature °C

FIGURE 9.3 Skin temperature largely follows skin blood flow. More heat is formed with moderately heavy work (thick line), which stimulates opening of the contracted vessels in the finger and warms it up. Light work (thin line), which means lower heat production, cannot stimulate opening of the vessels.

and feet therefore easily get cold in cold conditions (Figure 9.3). If the cooling continues, the body's other defense is to increase heat production by causing Bert to shiver (see Fact Box 9.3). Shivering is stimulated by a reduction in skin temperature and the resultant reduction in core temperature (core temperature represents 67–80% of the driving force behind shivering). Shivering therefore occurs in principle only when one is generally motionless.

9.5.1.2 Work

9.5.1.2.1 Energy Metabolism and Work Ability

The energy metabolism is not noticeably higher in the cold at a certain level of muscle work under other similar conditions, unless body temperature is reduced a great deal and/or shivering begins. Under circumstances like these, energy metabolism increases (see Fact Box 9.3). On the other hand, somewhat more energy may be metabolized in cold than at room temperature when Bert moves a great deal. One of the reasons for this is that he wears more clothes in the cold, which weigh more and in part restrict his body movements, so that more energy is needed to move (see Fact Box 9.3). In other cases, it might be because he wants to move more in the cold to keep warm.

The physical work ability can be limited in the cold, partly resulting from the fact that his joints and muscles are chilled. There will also be narrower margins up to Bert's maximum possible work output, as heavy winter clothes make for extra weight.

FACT BOX 9.3

- At work, energy metabolism increases by ~3% per extra kilogram of clothing.
- Shivering may increase energy consumption and heat production by up to approximately five times the rest value. When shivering, the skeletal muscles contract intermittently, thereby creating heat, which can warm up the body. The greater energy requirements in shivering are provided for by an increase in the glucose release from the store of glycogen in the liver.

9.5.1.2.2 Circulation and Work Ability

In the cold, the sympathetic nerves are activated, releasing the transmitter substance norepinephrine. This stimulates a contraction of vessels in the skin (see Figure 9.4) via adrenergic receptors (muscle tone in the vessel walls). Particularly significant for temperature regulation (both in heat and cold) are vessel structures called *arteriovenous anastomoses*, AVAs. AVAs form a direct link between arterioles and venules and can constrict or open, thereby substantially altering the blood flow in the intermediate capillaries. In fingers and toes, the ears and the nose, where the blood flow varies most, there are a great number of AVAs. As the blood flow can be cut off particularly effectively there, these parts of the body are also particularly vulnerable to cold. In warming up, the blood flow increases quickly again.

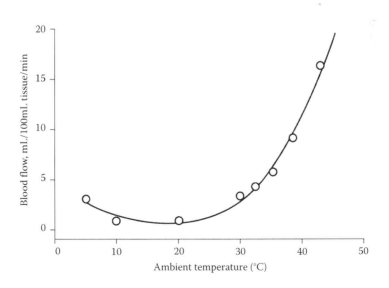

FIGURE 9.4 The dependence of blood flow on ambient temperature measured in the hand. (Data from Brown, GM. and J. Page. 1952. *J Appl Physiol.* 5:221–227.)

At skin temperature <31°C, the vessels of the skin are at maximum contraction. Blood flow in the skin is therefore low when sitting down or standing up. In the extremities the blood flow is redirected to deep veins and arteries, which function as a heat exchanger, so that heat is conserved.

A reduction in surface circulation of warm blood leads to a reduction in heat transfer from the skin. There are two reasons for less heat to be transferred. On the one hand, the temperature difference between the body surface and the surroundings is lower when the skin becomes cooler; on the contrary, the heat conduction capacity of skin with minimal circulation is lower than in tissue that is perfused with blood. Insulation of the tissue which is not perfused increases two to three times. Note that the blood flow to the head is not noticeably affected by the cold, as the blood vessels there have very few sympathetic nerve fibres. This results in a large proportion of the heat leaving the body being emitted from the head, if it is not protected.

The increase in central blood volume means that blood pressure increases. The body attempts to compensate for the increase in blood pressure. This is done by lowering the heart rate and reducing the heart's stroke volume, and through an increase in kidney filtration (more urine is produced). This is presumably the reason why we often need to urinate in cold weather. More fluid also "leaks" to the space between the cells (interstitial compartment). The proportion of blood volume excluding blood cells (plasma volume) may decrease by up to 15%. The blood therefore becomes more viscous and sluggish, which increases the risk of local frostbite and strain on the heart (see also Section 9.5.2).

9.5.1.2.3   The Musculoskeletal System

Cooling affects endurance, strength, power generation, speed, and coordination of movements. The speed of conductivity falls in nerves (by 1–2 m/s per degree) making reflexes and movements slower and less precise. The resistance of tissues can also deteriorate, with a greater risk of damage in muscle work in cold or with sudden overload.

Dynamic work ability decreases in the cold. Power generation falls ~3–8% for each degree of reduction in muscle temperature in short-term dynamic work. Moreover, isometric force development (in maximum voluntary contraction) is lower when the muscle is cold. Endurance in static work, however, has an optimum at around 27°C muscle temperature and falls both above and below this. In the cold, more motor units have to be recruited to achieve the same force as at room temperature, or else the active units have to increase their force (see Chapter 6, Section 6.6). Muscle fatigue therefore occurs earlier.

In most jobs, a good ability to work with the hands is required. Feeling in the fingers, which is very important in precision work, starts to drop off even at a skin temperature of 32°C (which is normal temperature) and falls further when cooled down to 7–8°C, when the nerve conductivity ceases. Mobility in the fingers begins to diminish at around 21°C (skin temperature) and becomes severely limited at about 16°C.

Apart from the physical work ability, mental functions are also affected by cooling, which produces uneasiness and may distract the individual at work.

9.5.1.2.4 Breathing

Breathing in cold air and exposing the face to cold air may impair physical work ability and produce disorders. A running nose is a common reaction to cold, which can be annoying and distracting.

Cold air is dry, containing only small amounts of water—just a few grams per kilogram of air. Warm air can contain considerably more water. The mucous membranes in the nose and airways humidify the dry air so that it becomes saturated with water vapor. In physically heavy work, when large volumes of cold air are exhaled through the mouth, there is a risk that the mucous membranes in the airways dry out and become irritated. See Section 9.5.2.

9.5.1.2.5 What Happens in the Cells?

The effect of exposure to cold on human cells has not been well researched. It has been observed that the lipid composition in the cell membrane changes, so that its function can be maintained, and that RNA synthesis, protein synthesis, and cell growth decrease. The protective so-called *cold-shock proteins* are formed when the cells are exposed to cold, and so they can better withstand the strain.

9.5.1.2.6 Cold Acclimatization

Acclimatization to cold in the long term can occur locally through vascular constriction in peripheral parts of the body not being activated in the cold. This means that the nerve–muscle function in hand/feet can be maintained, but at the same time means a greater heat transfer, which may cool down the body if exposure is long term. After a month of working outdoors, Bert's hands keep warm for much longer than at the beginning of the winter, but not all of his colleagues are affected in this way. It is very individual. It is still uncertain whether acclimatization of the whole body occurs in the cold. Psychological adaptation probably occurs, which means that the experience and the unpleasantness of cold declines after regular exposure.

9.5.2 RISKS IN COLD

By constricting AVAs in the skin, the body can maintain the heat balance if exposure is short term or mild, but at the cost of comfort. When the skin temperature falls, it first feels unpleasant, then a feeling of cold and pain follows. If the hands get cold, the risk of accidents in manual work increases, as muscle capacity falls off and coordination of movements is considerably impaired.

If the cold load is considerable, then vascular contraction is not enough to maintain the inner body temperature. Hypothermia, which implies that the body temperature falls below 35°C, is however unusual at work. Most cases of hypothermia have been among people who are intoxicated, have a psychological illness or dementia, or people who have had accidents or gotten lost. It is, however, important to be aware that certain pharmaceuticals, alcohol, and nicotine affect temperature regulation. The most important risk factors for *hypothermia and injuries* are shown in Fact Box 9.4.

FACT BOX 9.4

RISK FACTORS FOR HYPOTHERMIA AND INJURIES

• Wind	
• Handling cold metal objects and fluids	• Increases chilling and the risk of frostbite
• Inadequate or wet clothing	
• Low physical activity	• Low heat production, easy to become chilled
• Illness	• Increases sensitivity
• Particularly susceptible individuals	
• Ice and snow, darkness	• Makes work more difficult, risk of slipping

Frostbite occurs when the temperature of the tissues falls below 0°C. Ice crystals, which shred the cell membrane, are formed in the fluid between the cells, and the cells dry out (through osmosis). Frostbite can be severe, and in the worst cases results in tissue death, which requires amputation or may lead to increased sensitivity to cold and impaired sensitivity long after the injury has healed.

When the face is exposed to cold wind (or cold water), as when Bert is working outdoors in winter, a reflexogenic reduction in heart rate (via parasympaticus) occurs at the same time as an increase in blood pressure activated by the sympathetic system. This can cause problems for Bert, who has high blood pressure, and for people with other cardiovascular diseases, as it means greater strain on the heart.

Respiratory disorders can also be triggered by cold on the face and body. The bronchial tubes in the airways contract owing to stimulation of nerves under sympathetic control (β-adrenergic receptors). In extreme cold, heavy work with high lung ventilation may imply a risk of damage to the airways. In strenuous work in cold conditions, 4–20% of healthy individuals experience asthma-like symptoms. There are indications that heavy breathing of cold air can lead to inflammation of the lower airways. The risks increase at air temperatures below –20°C. Disorders of the musculoskeletal system are more common at cold workplaces than at those at room temperature, and seem to increase with exposure time. Bert's problems with his knees are presumably linked to his exposure to cold. Table 9.1 lists medical disorders that may arise from or be exacerbated by cold, and illnesses that may involve problems when working in the cold.

There are considerable individual differences in tolerance to cold, which have to be taken into account when making a risk assessment. Particular risk groups in working life are individuals:

- Who have previously been susceptible to cold or who have had frostbite.
- Who are inexperienced in working in the cold.
- With certain chronic illnesses (see above) and/or who are taking medicine which affects temperature regulation.
- With an ongoing infection and fever.

TABLE 9.1
Health Problems and Disorders with Cold Strain

Cardiovascular diseases	Ischemic heart disease (coronary disease, resulting in oxygen deficiency in the heart), high blood pressure (hypertension).
Cold—allergy	Allergic reaction with swelling and rash on the skin after reheating cold skin. Also general symptoms, such as headache, breathing difficulties (dyspnea), tachycardia (palpitations), and allergic shock occur.
Diabetes	In diabetes, nerve function and blood circulation are often impaired peripherally. This exacerbates the hypothermia.
Respiratory diseases	A large proportion of people with asthma and chronic obstructive lung disease experience disorders in the cold. The symptoms often arise in connection with exertion. People with chronic bronchitis also sometimes have asthma-like symptoms in the cold.
Raynaud's syndrome/white fingers	Impaired circulation in the fingers. The fingers go pale because the circulation is cut off. The syndrome is often the result of a vibration injury, but can also be an innate overreactivity on the part of the vessels.
Skin diseases	Psoriasis and various forms of skin complaints, which involve damage to dermal layers, may increase heat transfer from skin and thereby the risk of frostbite. Some types of dermatitis can these be exacerbated by cold, for example, atopic dermatitis.
Endocrine diseases	Deficient pituitary function and hypothyroidism (lack of the hormone thyroxine). The hormones are necessary for normal heat production and metabolism.
Diseases of the musculoskeletal system	Tendinitis, lumbar spine conditions, pain in shoulders and knees. Repetitive wrist movements in the cold increase the risk of carpal tunnel syndrome

9.6   HEAT

9.6.1   Physiological Responses to Heat and the Effect on Work Ability and Health

9.6.1.1   Rest

When Bert is exposed to heat at rest, his body protects itself against overheating through a major proportion of the blood flow being diverted from central parts of the body to the skin, mucous membranes, and peripheral parts of the body (Figure 9.4), which get warmer. In physical labour, large parts of the blood, as usual, go to the working muscles. If the heating continues, Bert begins to sweat to give off more heat.

9.6.1.2   Work

9.6.1.2.1   Energy Metabolism

Energy metabolism is in certain cases somewhat higher during work in the heat than at lower temperatures. When Bert is doing carpentry, his energy metabolism increases, and thereby the temperature in his muscles and the rest of his body. When he is doing heavy work, heat production is compensated for by heat transfer, which involves stress on the body even at normal temperatures. Hot working days also further limit how heavy his work can be, and how long he can work without risk to his health. The increase in muscle temperature also increases the degradation of glycogen in the muscle, reduces fat oxidation, and increases the accumulation of lactic acid (lactate). In heavy physical work, the release of adrenaline is increased (in the heat, around twice as much), which also increases the degradation of muscle glycogen.

9.6.1.2.2   Circulation

Rerouting of some of the blood volume that the heart pumps out every minute (cardiac output) from the internal organs to the muscles, skin, and peripheral parts of the body is carried out through stimulating the autonomous nervous system. In the heat this leads to the dilation of the vessels in the skin (relaxation of the musculature/reduction in muscle tension in the walls of the vessels) and an opening of arteriovenous anastomoses (see Section 9.5.1 on blood flow in the cold). In order to compensate for the greater volume of the vessels, the cardiac output increases through increasing the heart's contractive force and rate. The heart rate is thus a relatively simple measure of circulation load and heat load, which can be used in assessing physiological strain in the heat. Physical training to improve the heart's ability to work therefore provides much better preconditions for working in the heat. Bert, who is no longer so fit, therefore becomes more tired than before when doing heavy work in the heat.

9.6.1.2.3   Sweating and Fluid Balance

At rest, the body emits ~30 g (0.3 dL) of fluid per hour through the skin, the mucous membranes, and the lungs in the form of water vapor (also called *perspiratio insensibilis*) even when not sweating. The ability to sweat is the most important human attribute for tolerating heat. The body has between 2 million and 5 million sweat glands. We normally sweat 0.5–1.5 L/h in the heat, but in extreme conditions sweating may be as high as 3 L/h. When sweat evaporates (vaporizes), energy as heat is lost from the skin and this makes possible the removal of the heat carried there via the blood (Fact Box 9.5). There is a considerable individual variation in the sweating function, which depends on things such as hereditary factors, the level of physical training, and heat acclimatization. Both training and habituation (see *Acclimatization*, Section 9.6.2) increase the production of sweat and thereby cooling the body.

An increase in central and peripheral temperature (central temperature is more important than skin temperature) activates the sweat glands. The signals go out from the CNS to sympathetic postganglionic nerves (nerves that have one end in the mass of nerve cell bodies, the ganglia, outside the spinal cord) which stimulate the production of sweat. Peripheral thermoreceptors can modify the response.

FACT BOX 9.5

About 2430 kJ of thermal energy is used to evaporate 1 L of sweat.

9.6.1.2.4   The Musculoskeletal System

At higher muscle temperature the speed of the contraction increases. Maximum force development in the muscle seems to be little affected, however, by muscle temperatures between 25°C and 40°C. Heat load leads to more rapid fatigue and lower endurance both in dynamic and static work. One of the causes may be that central stimulation of the motor units decreases with heat load. The aerobic energy delivery declines with physical work in the heat, and lactate is more easily formed.

9.6.1.2.5   How is the Work Ability Affected?

The first noticeable effects of heat load, apart from the feeling of overheating, are psychologically conditioned disorders, such as discomfort and irritation. Both discomfort and heating can affect mental functions, such as attention, and the ability to carry out complex tasks. The effect is a gradual one, and eventually performance capacity becomes impaired. The risk of missing important information, for example, traffic signals when Bert is driving his car home from work on a hot day, increases with heat load. The extent to which different individuals are affected varies, and depends on how quickly heat is stored in the body, the type of task, and work skills.

If Bert does not have sufficient fluid intake during the working day, sweating leads to dehydration. The fluid deficit impairs his physical work ability, all the more so the greater the deficit. A fluid deficit of, for example, 1.6 L in an individual weighing 80 kg (i.e., to say 2% of body weight) suffers impairment in endurance of 10–20%. As sweat contains salt, fluid loss means that salt is lost at the same time. With major fluid loss, the salt has to be compensated for with things like energy drinks or extra salt. It should be noted that a normal diet often contains an excess of salt.

9.6.1.2.6   What Happens in the Cells?

When the temperature exceeds ~40°C in the muscle cells, permeability of the mitochondrial membrane and a number of metabolic changes affecting the function of the muscle presumably occur. For example, oxidative phosphorylation in the mitochondria, which provides energy-rich molecules for things like muscle work, decreases. The so-called heat-shock proteins (HSP) are formed with heat stress. They protect the DNA of the cell and other proteins from being destroyed. HSP is also formed as a result of other types of stress, such as oxygen deficiency and glucose deficiency in the cells. Knowledge of this subject area is still very limited.

9.6.2   ACCLIMATIZATION TO HEAT AND PERFORMANCE IN THE HEAT

By frequently being or working in the heat, adaptation of the entire system occurs, which means that the sweating capacity and regulation of blood flow are improved. This process is called heat acclimatization. The full effect is achieved after 7–10 days of regular exposure to heat. The majority of the positive effect is achieved as

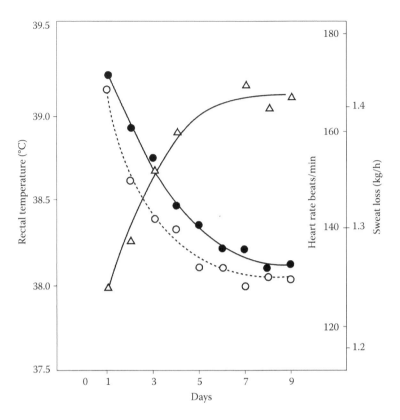

FIGURE 9.5 Physiological effects of 9 days' heat acclimatization on core (rectal) temperature, sweating, and heart rate. Triangles = sweat loss in kg per hour; hollow circles = rectal temperature in°C; solid circles = heart rate in beats per minute. (Modified from Lind, AR. and DE. Bass. 1963. *Fed. Proc.* 22:704–708. With permission.)

soon as 3–4 days. Physical exercise also results in similar physiological adaptations as in heat training. Heat training is, by analogy with physical exercise, more efficient the longer and hotter the daily dose. The effect of the training is a very marked decrease in the feeling of exertion and physiological strain at work.

After heat acclimatization at any given level of physical work, the heart rate is lower, the stroke volume and skin blood flow are greater, and heart activity and sweating more efficient (quicker activation of sweat glands and activation of more sweat glands, greater production of sweat), all of which leads to lower body temperature (Figure 9.5). As mentioned earlier, the distribution of blood pumped from the heart away from internal organs to the muscles and skin alters during work. After acclimatization, blood flow to the skin increases while maintaining circulation in the inner organs. The plasma volume increases as a result of the increase in the amount of protein in the plasma through synthesis, recycling, and fluid leaking in from the space outside the vessel. Total loss of heat acclimatization takes 3–4 weeks.

9.6.3   RISKS WITH HEAT

Heat balance can be maintained normally through the vessels in the skin dilating and as a result of sweating. If the ambient temperature (and humidity) is too high in moderately heavy work, for example, at an air temperature exceeding 25°C, these protective reactions are not enough, and body temperature increases. The greatest medical risks of working in the heat are overheating of the body, dehydration/fluid deficit, and burns. The effect on mental functions can have serious consequences with mistakes leading to accidents; for example, drivers missing traffic lights and causing a traffic accident. Several factors in the environment produce risks of heat strain or injuries:

- High air temperature (or water temperature when diving)
- High humidity
- Solar radiation or heat radiation
- Hot surfaces and fluids
- Tight clothes
- High work rate, heavy physical work

Combinations of several of these factors increase the risk. High humidity in combination with high air temperature, for example, mean that not less sweat evaporates from the skin over a given period of time, which increases the rate at which the body heats up. In fire fighting with breathing apparatus, to take an extreme example, fire fighters are exposed to all of the above-mentioned risk factors.

The high demands for blood supply to both the working muscles and skin at the same time can lead to overload, exhaustion, and collapse (see more below).

Intensive sweating and drinking too little can lead to dehydration and salt deficiency (see Fact Box 9.6). In dehydration, the plasma volume, and the stroke volume and minute volume of the heart decrease (this is also the result of overheating). Fainting may occur as a result of fluid deficit (circulation to the brain decreases), and muscle cramps as a result of major losses of salt and minerals and muscle fatigue. If heat syncope or other heat problems occur, the correct acute treatment should be given, and possible hospital treatment should follow.

FACT BOX 9.6

Sweat contains 0.5–1.0% minerals and organic substances, primarily NaCl (salt), urea, lactate, and potassium ions.

Another risk in heat strain is too excessive a sugar concentration in the blood (hyperglycemia), as the glucose uptake does not increase in the cells, despite the fact that glycogen degradation increases in heat (see p. 250).

Burns, which can occur in contact with hot surfaces and fluids, require rapid cooling and often medical attention.

Most people cope with reasonable heat load and may, as mentioned above, train their tolerance. The physiological capacity in work in the heat, however, varies a great deal between individuals. A very small group of people are heat intolerant, and suffer severely in hot environments. A large number of illnesses, obesity, or a generally fragile state of health makes for greater sensitivity to heat load. Alcohol, caffeine, nicotine, and other drugs as well as certain pharmaceuticals impair heat tolerance by affecting the CNS or peripheral functions, for example, vascular contraction and relaxation. For instance, antihistamines (for allergies) impair heat transfer, both through their influence on the hypothalamus and on the sweat glands.

Apart from climate exposure, there are several individual factors that increase the risk of heat strain:

- Pregnancy (higher heat production)
- Small body mass (heats up quickly)
- Obesity (circulation already loaded)
- Age (less tolerance with increased age)
- Cardiovascular disease (circulation functions less well)
- Diabetes (poor peripheral circulation)
- Skin complaints, for example, scleroderma (impaired peripheral circulation)

Overheating of the body can be dangerous (Table 9.2). It is therefore important that both the person working in the heat and the supervisor are aware of early signs (considerable discomfort and a feeling of being hot, a paradoxical experience of cold, in combination with thirst, headache, a feeling of being sick, and impaired concentration). Work should be stopped immediately and cooling measures started straight away. In cases of heat stroke, body temperature continues to rise uncontrollably if nothing is done. Together with burns these are the most serious conditions in the heat, and they require immediate medical intervention.

TABLE 9.2
Medical Consequences of Excessive Heat Load

Dehydration	Leads to impaired sweating function in circulation, and may lead to headache, dizziness, and fainting.
Edema	Swelling of hands, feet, and ankles as a result of salt and fluid deficit.
Heat rash	Characterized by increased histamine response leading to eczema and respiratory disorders.
Heat cramp	Muscle cramp as a result of salt deficit, fluid deficit, and muscle fatigue.
Heat exhaustion (resulting from a fall in blood pressure) and muscle cramp	Sometimes in combination with raised body temperature, often resulting from fluid deficit leading to weakness, a feeling of sickness, dizziness, disorientation, and sometimes fainting.
Heat syncope, heat stroke (resulting from high body temperature)	Sweating ceases, serious CNS impact with symptoms as for heat exhaustion as well as initially high heart rate and later weak pulse and renal failure

9.7 INDOOR CLIMATE

Indoor climate normally implies no great strains on the body in the way that a more extreme climate does. Despite this, complaints are often made about the climate particularly in offices. Common concerns are that it is draughty and too cold. One explanation for this is that office work normally involves a great deal of sitting/standing, where the individual has low heat production, and is therefore sensitive to small variations in the climate of the surroundings. Another contributory factor is that people wear light clothes, which do not protect all parts of the body equally, for example, the neck or the legs. Movements of the air are felt as draughts when they exceed around 0.15 m/s (0.3 miles/h) (depending to a certain extent on the air temperature and degree of turbulence). Cooling that only affects certain parts of the body contributes to the total climate experience, even if central body temperature is not particularly affected. Low temperature outdoors means that poorly insulated walls and windows lead to cold and radiation draught indoors. Radiation draught means an experience of draught which is not caused by high air velocity, but rather by heat being lost through radiation from part of the body to a cold surface. Particularly cold indoor climates, for example, cold stores in the food industry, are dealt with in the section on cold above.

In summer, solar radiation means that it becomes too hot indoors, for example, in buildings with large glass windows. At the beginning of a heat wave, it is usual for many people working in offices to complain about the heat. Unfamiliarity with heat, both psychological and physiological, means that many people experience the first days as trying. Gradually, familiarization occurs, and the discomfort and strain decline. It is frequently possible to find the causes of the complaints in the physical work environment. Perhaps the ventilation plant is not dimensioned correctly for cold air. Staff that have external contacts at work usually have some kind of formal dress or uniform and cannot therefore always adapt their clothing to the heat. There may be an opportunity to wear a lighter variant of the formal uniform which is acceptable, which is a negotiating issue between the employees and employer. For these reasons, it is a good idea to distinguish between the thermal climate and other environmental factors when presenting or discussing factors in the indoor environment, as sometimes the concept of indoor climate also comprises air quality, electrical and magnetic factors, as well as light and sound factors. Only thermal factors are dealt with in detail in this book.

High air temperature does, however, contribute to the fact that air quality is experienced as poor, and that the air feels stale. Sometimes discontent at other conditions of the working environment, for example, poor psychosocial environment, are projected onto the thermal climate and air quality.

9.8 GENDER ASPECTS OF WORK IN AN EXTREME CLIMATE

Some factors that are relevant to discussions relating to the significance of gender for climate impact are:

- Body mass
- Surface/volume ratio

- Subcutaneous fat
- Physical capacity (which influences heat production)
- Perspiration (which influences heat transfer)

The first three aspects of climate mentioned above are connected with body dimensions and build, and are not therefore gender factors in themselves. Cooling and heating of a body with less mass occur more rapidly than one with more mass. They also occur more rapidly if the surface is greater in relation to the volume is greater, as heat is emitted from the body surface. As a group, women have both less absolute body mass and a lighter build (smaller diameter of the skeleton and muscles). Individuals with a small body mass and slimmer build cannot normally stay as long in a hot or cold climate as a bigger and more powerfully built person. In physically heavier work, one's own heat production also influences how great heating or cooling is. A high physical capacity being exploited means a high heat production, which counteracts cooling in cold locations and increases the heat load in hot locations.

Women generally have more subcutaneous fat than men. In cold, when the vessels are contracted peripherally, the amount of subcutaneous fat has a certain significance for cooling the body, as fat is a good insulator (has a low heat conduction capacity). It is, however, primarily in water that the effects of the greater insulation become noticeable, for example, in diving work.

Sweating, which is necessary to cool the body in great heat, differs somewhat between the genders, even if one takes body size into account. Women sweat less than men with exertion and/or heat exposure. There are differences in both the amount of sweat and how quickly sweating begins. In situations where the opportunities for evaporation are limited, as where humidity is very high, or tight protective clothing is being worn, the difference has little significance. Then body size is decisive for how much heat can be stored before body temperature gets too high. All in all, the theoretical differences between the genders from a climate point of view do not have a great practical significance in normal working conditions. Variations *within* the entire group of men and women as regards body dimensions, experience, and physiological reactions presumably contribute to greater differences in climatic effects than the differences *between* the genders.

9.9   AGE ASPECTS OF WORK IN AN EXTREME CLIMATE

Increasing age is often associated with deteriorating physiological functions, including temperature regulation and aerobic capacity (see Chapter 2). It is, however, difficult to distinguish between what directly results from chronological age and what derives from the change in lifestyle that often accompanies age (e.g., poorer physical work capacity and increased sitting still). With age, the sensitivity in the temperature-regulating structures and organs (e.g., sweat glands and blood vessels) decreases and structural changes occur in the skin. This means that "protective" reactions to heat and cold, such as constriction or dilation of the vessels in the skin, begin later in older people than in younger ones, and sweat production is lower in older people than in younger ones. Tolerance of heat and cold may therefore decline with age. One should be aware that a great deal of the knowledge about age and climate is based on studies of individuals who have progressed beyond the age of active working life. Apart from the normal physiological changes already mentioned, the relatively common illness-induced impairments in aerobic capacity occur, for example, in ischemic heart disease.

To sum up, the observed differences directly related to age are relatively minor, but if we take into account the physiological consequences, and lifestyle changes in advancing years, the tolerance of climate effects are somewhat lower in older individuals than in younger ones.

9.10 MEASUREMENT AND RATING OF THERMAL CLIMATE

In order to assess the effect of the thermal climate, the following are used, listed by increasing degree of difficulty:

- Subjective ratings
- Measurements of physical factors
- Physiological measurements

and combinations of these.

9.10.1 Scales for Rating Climate Experience

By asking employers to assess their experiences by using standardized rating scales [ISO 10551 1995], we gain an understanding of how difficult the climate problem is regarded as being. It is simple from the viewpoint of measurement technology but only provides a rough measure of assessment. One example of a commonly used rating scale for heat and cold can be seen in Figure 9.6. To discover causes and to make an overall assessment based on detailed information, we need to measure the physical environment and evaluate and measure workload and the properties of clothing.

How do you rate the thermal
sensation of your body?

+3	Hot
+2	Warm
+1	Slightly warm
0	Neutral
−1	Slighty cool
−2	Cool
−3	Cold

FIGURE 9.6 Rating scale for thermal sensation. This can also be used for individual parts of the body, for example, hands and feet. (From ISO 10551. 1995. *Ergonomics of the Thermal Environment—Assessment of the Influence of the Thermal Environment Using Subjective Judgement Scales.* International Organization of Standardization, Geneva.)

9.10.2 MEASURING PHYSICAL FACTORS

To be able to assess the effect of climate on the body, physical measurements need to be carried out. The measurements should properly comprise the following:

- Air temperature
- Radiant temperature
- Mean radiant temperature
- Air velocity
- Air humidity
- Work rate (work load, heat production)
- Clothing insulation and water vapor resistance.

Air temperature can be measured quite simply with a screened thermometer. Radiant temperature requires special sensors and is measured in six directions in the room (ceiling, floor, and all the walls). Air velocity can be measured mechanically with a turbine wheel, but indoors electronic instruments are preferable, as they are more sensitive. Humidity can be measured using a modified thermometer or with sophisticated air humidity sensors. For certain assessments, special types of temperature have to be measured. The sensors are specially designed and are used in particular situations, for example, globe temperature and natural wet bulb temperature for Wet Bulb Globe Temperatures, an index for assessing heat load. Detailed descriptions of measuring instruments for climatic factors are to be found in the standard ISO 7726 [1998].

To measure oxygen consumption/workload/heat production and the thermal properties of clothing is more complicated than the other factors, as these require particular methods and measuring instruments which are not normally readily available. They are therefore assessed most readily with the help of indirect methods and estimates.

Assessment of metabolic rate and calculation of heat production can be carried out with the help of formulae and tables in the standard ISO 8996 [2004]. Measurement of work rate is described in Chapter 2. From oxygen consumption during the job in question, heat production can be calculated or approximated. Heat production (H) = metabolic rate (M)—the mechanical work rate (W) performed (see Section 9.4).

For many jobs involving static muscle work mechanical efficiency is so low (see Chapter 2) that heat production can be equated with the total metabolic rate, that is to say, $H \sim M$.

The calculation of metabolic rate M (W/m^2) is:

$$M = EE \times V_{O_2} \, L/h \, (W) \tag{9.1}$$

in which EE = energetic equivalent in Watt-hours per liter of oxygen and V_{O_2} is the oxygen consumption in liters of oxygen per hour. The conversion coefficient for calculating metabolic rate is assumed to be 20.6 kJ/L of oxygen [ISO 8996 2004].

The thermal properties of clothing (insulation and water vapor resistance) are measured on what is called a thermal manikin. Insulation and water vapor resistance can also be estimated with the help of tables to be found in the standard ISO 9920 [2007].

CALCULATIONS 9.1

When Bert is hammering nails, he uses 0.8 L of oxygen per minute. Metabolism yields 20.6 kJ/L of oxygen, which gives $0.8 \times 20\ 600\ J/60\ s = 275\ W$. The efficiency is ~15% for hammering nails and the rest, that is to say $275\ W \times 0.85 = 233\ W$, is therefore converted into heat.

9.10.3 PHYSIOLOGICAL MEASUREMENTS

It is also possible to measure the effect of the climate on the individual by carrying out physiological measurements, that is to say by measuring heart rate (see also Chapter 2), fluid loss, body temperature, and skin temperature. It is important that the measurements are carried out in a standardized manner. The methods are described in international standard ISO 9886 [2004]. Physiological measurements may also need to be carried out in medical monitoring and checks on individuals who are assessed as being at great risk in hot and cold environments ISO 12894 [2001]. This can help to discover early signs of ill health, to protect particularly sensitive individuals, and to prevent accidents conditioned by certain medical complaints, for example, impaired judgement or ability to react on the part of drivers.

9.10.4 METHODS FOR ASSESSING THE RISK OF HYPOTHERMIA

It is not acceptable for the body's core temperature to drop during the working day. Thermal comfort implies that the climate is felt to be pleasant. Comfort should be the aim, of course, but it is not always possible to achieve this in an extremely cold climate. There are tried and tested methods that can be used for assessing how cold affects the individual. Those methods described here are the ones most frequently used for assessing cold strain, and stress.

9.10.4.1   Insulation REQuired Index, Insulation Needs

The Insulation REQuired Index (IREQ) is a method used for assessing the risk of cooling of the whole body. This is based on a calculation of the heat balance and environmental factors affecting the heat balance. The result of the calculation comprises a value for the insulation on the part of the clothing system needed to maintain the heat balance. Good insulation properties, for example, in the material Styrofoam, mean that little heat escapes from the material, in this case clothes. IREQ is expressed by the unit $m^2 \times °C/W$ or clo (1 clo = 0.155 $m^2 \times °C/W$ and corresponds to the insulation in normal clothing indoors in winter).

For work at a given level of physical activity in a given thermal environment, certain clothing insulation is required to counteract hypothermia of the body core and to provide comfort. The connection between these factors is illustrated by the

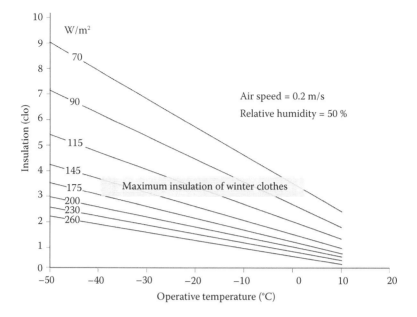

FIGURE 9.7   Requirement for clothing insulation to avoid hypothermia in various combinations of work rate (lines marked W/m²) and operative temperature (*x*-axis). Moderately heavy work (175 W/m² body area) requires an insulation of ~2 clo at −20°C. 1 clo corresponds to normal indoor clothing. The practical upper limit for insulation with current clothing systems is 3–4 clo (shaded in the figure). Above this limit, working time has to be restricted to avoid hypothermia (see Section 9.10.4). Operative temperature is an overall appraisal of air and radiant temperature. Air temperature = operative temperature in a room where the ceiling, walls, and floor are at the same temperature as the air. (Modified from Gavhed, D. and I. Holmér. 2006. *The Thermal Climate at the Workplace.* Arbetslivsrapport 2006:2, National Institute for Working Life, Stockholm. (In Swedish).)

diagrams in Figure 9.7. The diagram can be used to predict clothing needs, to choose suitable working clothes, and to assess the risk of hypothermia in the body core during a working day.

Despite the fact that Bert has adequate clothes to protect him from the cold, he cannot work an entire day in certain climatic conditions. In work while standing in an air temperature below approximately −10°C, there are no work clothes that can protect sufficiently against hypothermia of the body. Working hours then have to be limited. The longest suitable shift (DLE, Duration Limited Exposure) can be calculated. There are diagrams, tables, and computer programmes for calculating IREQ and DLE in the standard ISO 11079 [2007].

9.10.4.2   Wind Chill Index

Even when Bert has established a heat balance, he risks frostbite in severe cold. Bare, unprotected skin is most vulnerable. Often the face and ears are not protected, and sometimes work has to be carried out with bare hands even in the cold. Chill increases noticeably with wind. The Wind Chill Index, WCI, is used for assessing the risk of

TABLE 9.3
Wind Chill Index

Air Velocity
(m/s)   Air Temperature (°C)

	0	−5	−10	−15	−20	−25	−30	−35	−40
2	−1	−6	−11	−16	−21	−27	−32	−37	−42
5	−9	−15	−21	−28	−34	−40	−47	−53	−59
8	−13	−20	−27	−34	−41	−48	−55	−62	−69
16	−18	−26	−34	−42	−49	−57	−65	−73	−80
25	−20	−28	−36	−44	−52	−60	−69	−77	−85

Note: The figures show the temperature in °C which with no wind provides the same chilling effect (heat loss) in combination with a cold wind at a certain air temperature when bare skin is exposed to the cold wind. Can be used for assessing the risk of discomfort and frostbite. Shaded squares mean high risk of frostbite. Care should also be taken at a wind chill temperature of −21°C, as pain can occur and the risk of frostbite exists with longer exposures.

frostbite on bare skin. It cannot therefore be applied to parts of the body covered by clothing. The method is based on a calculation of the local heat transfer from the skin to the surroundings at a given air temperature and wind speed. As a help in the assessment, there is a table in which the risk levels are indicated (Table 9.3). Both IREQ and WCI are described in detail in the international standard ISO 11079 [2007].

Assessment of the risk of frostbite in contact with cold surfaces: Methods for assessing frostbite arising in contact with cold surfaces are described in the standard ISO 13732-3 [2005]. The risk is dependent on the surface temperature, duration of contact, the material, and its heat conduction capacity (e.g., wood has a low and metal high heat conduction capacity) and the pressure against the contact surface.

9.10.5 Methods for Assessing the Risk of Heat Strain

9.10.5.1 Wet Bulb Globe Temperature

Wet bulb globe temperature (WBGT) is a heat index for heat load, requiring relatively simple and straightforward measurements. It is calculated on the basis of measuring air humidity and globe temperature (a kind of average of air and radiant temperature, measured with a thermometer in a metal globe painted black). In addition, a rough classification of physical activity is required. WBGT is used to assess the effects of continuous or varying work in heat, and is particularly suitable for assessing environments where there is heat radiation.

The measured WBGT value is compared with reference values, which must not be exceeded. The reference values indicate a level of heat load that is acceptable and safe for almost all healthy individuals. This does not therefore apply to people who, for various reasons, have an impaired heat tolerance (see Section 9.6.3).

For most of the existing heat problems in working life, WBGT is a simple, practicable and sufficiently accurate assessment method. More about WBGT can be found in the standard ISO 7243 [1989].

On occasions when a more detailed and analytical assessment of the climate situation is required, or if the WBGT is exceeded, one might on occasion need to use PHS (see below).

9.10.5.2 Predicted Heat Strain

The predicted heat strain (PHS) method is based on a calculation of the heat balance so as not to exceed an acceptable limit for heat storage and fluid loss in a normal individual [ISO 7933 2004]. PHS requires more complicated measurements, calculations, and assessments than WBGT, and should only be carried out by specially trained staff.

Assessment for risk of burns from contact with hot objects: There are methods for assessing the risk of burns in contact with hot surfaces [ISO 13732-1 2006]. In the same way as in the case of contact with cold surfaces, the risk depends on more factors than merely temperature (see Section 9.6.3).

9.10.6 Methods for Assessing Discomfort from Heat and Cold in Indoor Climate

9.10.6.1 Predicted Mean Vote

Predicted mean vote (PMV) is a comfort index used for assessing thermal indoor climate (10–30°C) [ISO 7730 2005]. It is based on an empirical correlation between thermal sensation, the ambient climate, physical activity, and clothing (Figure 9.8). PMV is a value on the rating scale presented in Figure 9.6. The value 0 indicates that a group of individuals with the same combination of workload, clothing insulation and climate—in an office landscape, for example—would consider that it was neither hot nor cold. For each PMV value there is a predicted percentage dissatisfied (PPD) value, which can be used to predict the levels of demand placed on the workplace climate. It has been ascertained that, however good the climate, there will nevertheless always be around 5% of people who are dissatisfied because of the individual variations in experience.

9.11 REMEDIAL MEASURES FOR CLIMATE PROBLEMS

Climate problems, like other work environment problems, should be *prevented*. First, the causes of the problem should be eliminated if possible; second, the sources of the problem should be reduced, and finally, the individual should be protected with protective equipment and measures tailored to the individual. In outdoor work, it is of course a question of reducing exposure and protecting the individual from the wind and weather.

At the *social level*, laws and regulations provide guidelines for the consideration that should be shown. In the field of climate, the rules are in many cases quite general and do not go into detail. As definitive exposure limits for climate do not exist, a good knowledge of the effects of climate on working people is necessary for preventive

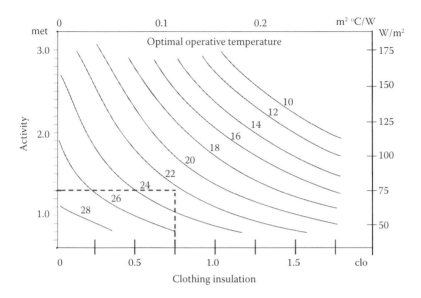

FIGURE 9.8 Recommended operative temperature when working at a certain work rate (Activity, met) with certain clothing insulation (Clothing insulation, clo). Operative temperature: see the caption of Figure 9.7. 1 met = 58 W/m², 1.0 clo = 0.155 m²°C/W. The broken line shows that at 1.2 met and 0.75 clo, 22°C is optimal. (Modified from Gavhed, D. and I. Holmér. 2006. *The Thermal Climate at the Workplace.* Arbetslivsrapport 2006:2, National Institute for Working Life, Stockholm. (In Swedish).)

work. The aim of the rules is to create a safe working environment for the promotion of good health, and thermal comfort is also the aim to the greatest possible extent. The employer has some leeway in achieving this and the way it is done may vary.

At the *management level*, risk assessment has to be carried out, and action plans for dealing with climate risks have to be produced as part of the systematic work on the working environment. In certain businesses, risks associated with climate occur only sporadically and for short periods or periodically (e.g., repairing the ovens in steel works).

Risk management should comprise:

- Introduction and training in risks in hot and cold work.
- Medical checks with special regard to risk groups.
- Good organization of tasks across the working day and the season, so that to the greatest extent the work is allocated to the time involving the least load from a climate viewpoint.
- Climate protection around building sites against wind in cold weather and solar radiation in warm weather.
- Supervision by workmates or management in high-risk work.

If the risks remain when the employer has carried out measures in the physical environment and organization of work, then measures will be required at an individual level, such as the improvement of clothes and breaks. Individuals can themselves, as far as possible, adapt their work rates (workload, apportionment of work/breaks) to climatic circumstances.

9.11.1 MEASURES IN THE COLD

At cold workplaces it is important, apart from the measures already mentioned, to arrange good opportunities for heating; warm facilities for taking breaks, infrared heat for warming hands, heated cabinets for drying clothes and regular breaks for warming up.

9.11.1.1 Clothes against the Cold

Clothes provide one of the most important opportunities for protecting the individual against hypothermia, and may often be the only factor in the environment which it is possible to vary to adapt to different climatic conditions at work. Insulation in clothes is important. The best insulator in clothes is stationary air. Apart from exploiting this fact, clothes systems against the cold should be based on the multi-layer principle, which means that people wear several layers of clothes one on top of the other. On the one hand, the different layers may have different functions, on the other, air that the body has warmed up can be enveloped between the layers and the insulation can thus be improved.

An efficiently functioning clothing system is built up according to the following principles:

- The clothes nearest to the body should fit closely around the body, so that the heated air next to the skin remains there. The material must not absorb moisture, because body moisture and sweat must not remain close to the skin. If sweating is not expected, it is possible to use absorbent materials.
- The next layer of clothing should insulate well, be flexible and keep its form well. In a very cold climate the intermediate layer may need to consist of two insulating layers. Common materials in the intermediate layer are wool and wool mixes as well as purely synthetic and insulating materials, such as fleece and fur fibre.
- The outer layer should be a shell garment which lets through water vapor yet is windproof and durable. Windproofing is particularly important if the garment is to be used outdoors or in cold stores or freezers, where fans cause considerable air movement, and headwinds cause cooling where industrial trucks are being used. In jobs where workers are exposed to the cold for a short time, extra clothing, such as a heat vest or jacket, is used on top of the ordinary working clothes.

Apart from the trunk, the arms and legs, the head, the feet, and the hands need to be well protected in the cold. A relatively large amount of heat is lost from the head, as the blood vessels in the crown and back of the head do not contract in the same way as in other peripheral parts of the body. Hands and feet need a great deal of insulation in relation to other parts of the body, as they have a large area compared with their mass and easily lose heat. It is also important of course to keep the body warm, because you have to be able to use your hands and feet in almost all work.

Clothing systems for the cold result in certain negative ergonomic consequences. Clothing systems comprise many garments that together add up to a relatively heavy weight. In this way, the range of motion and freedom of movement are limited, and the energy demand increases because of the weight. For this reason, it is best to choose light garments, which also have a well-thought-out functional design. Hoods, and to a certain extent head gear, limit the field of vision and the ability to hear so that communication with the surroundings is impaired, which is a considerable risk factor in many workplaces. All garments and possible protective equipment should also work together with each other, for example, gloves and cuffs or helmets and collars.

9.11.2 MEASURES IN THE HEAT

Heat storage in the body can be prevented largely in three different ways:

- Reducing the heat built up in physical work.

Examples of measures: reduce the total working time per hour with breaks or reduce the workload.

- Prevent heat input from the surroundings via direct radiation or hot air.

Examples of measures: screen off sources of radiant heat and use fans if the air temperature is lower than ~35°C (at higher temperatures the body receives heat and heat cannot be transfer through convection).

- Increase in heat transfer from the body, primarily through sweat evaporation.

Examples of measures: make evaporation easier by reducing the air humidity, using suitable clothing and drinks, conduct heat training.

It is important that clothes do not prevent heat transfer. They should therefore be permeable to air and water vapor and allow a large amount of air exchange through the clothes. In situations where extreme heat exposure cannot be avoided, protective clothing, and other protective equipment are required. Then the system that provides the best heat transfer should be chosen.

It is very important to have drinks on hand at the workplace. Prompt and regular consumption of water, corresponding to the fluid loss, counteracts the detrimental effects of dehydration to a great extent (Fact Box 9.7).

FACT BOX 9.7

Water is normally the best drink to replace fluid losses through sweating. Note that the experience of thirst is not a sufficient signal to cover the need. Maintaining body weight during the day is a measure of sufficient fluid compensation. A mug of water at 10–15°C every 20 min during physical work is a benchmark. This should, however, be adapted to the existing circumstances. In long-term and heavy sweating (more than an hour and more than 21/2 L), extra salt is needed, with normal salt intake otherwise. The salt content should not be <0.7 mg/mL in drinks for fluid compensation for sweating.

Coffee and other drinks containing caffeine and beer should be avoided, as they increase the excretion of urine. Drinks with a great deal of sugar should also be avoided.

9.12   WHAT THE LAW SAYS ABOUT WORK IN HEAT AND COLD

The laws are in general not detailed as regards climate exposure and climate load, but are merely frameworks for the physical work environment. The EU working environment directive states that the thermal climate *should be suitable and adapted* to the activity (Directive 89/654/EEC). Nor is there legislation in the United States that addresses the temperature of a workplace, unless one considers the General Duty Clause of the Occupational Safety and Health Act. According to Canadian regulations, the resulting thermal strain of the physical exposure is considered, for example, a worker's core body temperature should not exceed 38°C or fall below 36°C. For certain establishments, such as care facilities, special temperature limits apply. The laws usually prescribe that work hygiene conditions, as regards air, for example, should be satisfactory and that personal protective equipment should be used when adequate protection against ill health and accidents cannot be provided by other means.

There are therefore few exposure limits for the lowest temperature at which work can be carried out. Sometimes such exposure limits are written into local agreements. However, recommendations for prevention of thermal strain on workers exist in international guidelines and standards. For example, threshold-limited values (TLV) for heat exposure are provided in standard ISO 13732-1 [2006]. As they are based on how hard one is working and whether one is heat acclimatized, there are several exposure limits. In light work, for example, the WBGT should not exceed 30°C, while in heavy work it should not exceed 26°C.

PMV [ISO 7730 2005] provides *benchmarks* for the operative temperature indoors, 22°C ± 2°C for sedentary work in normal clothing, but provides no limit values. It is also pointed out in the standard that the workplace should be designed keeping in mind what suits different individuals.

The reason for the lack of exposure limits in laws on the working environment is that climate load (the "inner" exposure) varies depending on several environmental factors (air temperature, radiant temperature, air humidity, and air velocity), on workload and on the properties of the work clothes. As detailed exposure limits are not available, assessments of the risk discomfort and ill health have to be made in each individual case.

9.13 SUMMARY

Heat and cold increase the strain on most of the body's physiological systems. Body temperature is regulated by the nervous system to counteract overheating and hypothermia with the aid of changes in blood flow to the peripheral parts of the body. In the heat, sweating helps considerably in the cooling process. The effects of climate exposure can vary as to degree of severity, from discomfort to health problems. The individual's innate ability to adapt and physical training can affect the extent to which the climate impacts on human beings. It is important that the risks of climate exposure are assessed so that relevant preventive measures can be taken to ameliorate load and to maintain the individual's work ability. Assessment and measures can be supported by laws, regulations, and standards.

REFERENCES

Brown, GM. and J. Page. 1952. The effect of chronic exposure to cold on temperature and blood flow of the hand. *J Appl Physiol.* 5:221–227.

Directive 89/654/EEC—*Concerning the minimum safety and health requirements for the workplace.* European Agency for Safety and Health at Work. http://osha.europa.eu/sv/legislation/directives/workplaces-equipment-signs-personal-protective-equipment/osh-directives

Eurofound. 2007. *Fourth European Working Conditions Survey.* European Foundation for the Improvement of Living and Working Conditions. Office for Official Publications of the European Communities, Luxembourg.

Gavhed, D. and I. Holmér. 2006. *The Thermal Climate at the Workplace.* Arbetslivsrapport 2006:2, National Institute for Working Life, Stockholm (in Swedish).

ISO 7243. 1989. *Hot Environments—Estimation of the Heat stress on Working Man, Based on the WBGT-Index (Wet Bulb Globe Temperature).* International Organization of Standardization, Geneva.

ISO 7726. 1998. *Ergonomics of the Thermal Environment—Instruments for Measuring Physical Quantities.* International Organization of Standardization, Geneva.

ISO 7730. 2005. *Ergonomics of the Thermal Environment—Analytical Determination and Interpretation of Thermal Comfort Using Calculation of the PMV and PPD Indices and Local Thermal Comfort Criteria.* International Organization of Standardization, Geneva.

ISO 7933. 2004. *Ergonomics of the Thermal Environment—Analytical Determination and Interpretation of Heat Stress Using Calculation of the Predicted Heat Strain.* International Organization of Standardization, Geneva.

ISO 8996. 2004. *Ergonomics of the Thermal Environment—Determination of Metabolic Rate.* International Organization of Standardization, Geneva.

ISO 9886. 2004. *Ergonomics—Evaluation of Thermal Strain by Physiological Measurements.* International Organization of Standardization, Geneva.

ISO 9920. 2007. *Ergonomics of the Thermal Environment—Estimation of Thermal Insulation and Water Vapour Resistance of a Clothing Ensemble.* International Organization of Standardization, Geneva.

ISO 10551. 1995. *Ergonomics of the Thermal Environment—Assessment of the Influence of the Thermal Environment Using Subjective Judgement Scales.* International Organization of Standardization, Geneva.

ISO 11079. 2007. *Ergonomics of the Thermal Environment—Determination and Interpretation of Cold Stress When Using Required Clothing Insulation (IREQ) and Local Cooling Effects.* International Organization of Standardization, Geneva.

ISO 12894. 2001. *Ergonomics of the Thermal Environment—Medical Supervision of Individuals Exposed to Extreme Hot or Cold Environments.* International Organization of Standardization, Geneva.

ISO 13732-1. 2006. *Ergonomics of the Thermal Environment—Methods for the Assessment of Human Responses to Contact with Surfaces—Part 1: Hot Surfaces.* International Organization of Standardization, Geneva.

ISO 13732-3. 2005. *Ergonomics of the Thermal Environment—Methods for the Assessment of Human Responses to Contact with Surfaces—Part 3: Cold Surfaces.* International Organization of Standardization, Geneva.

Lind, AR. and DE. Bass. 1963. Optimal exposure time for development of acclimatization to heat. *Fed. Proc.* 22:704–708.

FURTHER READING

Åstrand, PO., K. Rodahl, HA. Dahl, and S. Stromme. 2003. *Textbook of Work Physiology-Physiological Bases of Exercise*, Chapter 13, Temperature regulation. Human Kinetics Canada, Windsor, Ontario.

Fregly, MJ. and CM. Blatteis (eds.). 1996. *Handbook of Physiology, Section 4: Environmental Physiology.* Vol. 1. Oxford University Press, New York, NY.

International Organization of Standardization, Geneva. International Ergonomics/thermal standards and standards for man-machine-interaction; http://www.iso.org

ISO 11399. 1995. Ergonomics of the thermal environment—Principles and application of relevant International Standards.

ISO 13731. 2001. Ergonomics of the thermal environment—Vocabulary and symbols.

ISO 14505-2. 2006. Ergonomics of the thermal environment—Evaluation of thermal environments in vehicles—Part 2: Determination of equivalent temperature.

ISO 14505-3. 2006. Ergonomics of the thermal environment—Evaluation of thermal environments in vehicles—Part 3: Evaluation of thermal comfort using human subjects.

ISO 15265. 2004. Ergonomics of the thermal environment—Risk assessment strategy for the prevention of stress or discomfort in thermal working conditions.

ISO 15743. 2008. Ergonomics of the thermal environment—Cold workplaces—Risk assessment and management.

ISO/TS 13732-2. 2001. Ergonomics of the thermal environment—Methods for the assessment of human responses to contact with surfaces—Part 2: Human contact with surfaces at moderate temperature.

ISO/TS 14415. 2005. Ergonomics of the thermal environment—Application of International Standards to people with special requirements.

ISO/TS 14505-1. 2007. Ergonomics of the thermal environment—Evaluation of thermal environments in vehicles—Part 1: Principles and methods for assessment of thermal stress.

Kroemer, KHE. and E. Grandjean. 1997. *Fitting the Task to the Human: A Textbook of Occupational Ergonomics*, Chapter 20, Indoor climate. Taylor & Francis, London.

Oksa, J., and H. Rintamäki. 1995. Dynamic work in cold. *Arctic Med Res.* 54(Suppl 2):29–31.

Parsons, KC. 2003. *Human Thermal Environments: The Effects of Hot, Moderate, and Cold Environments on Human Health, Comfort, and Performance.* Taylor & Francis, London.

Stellman, JM. (ed.). 1998. *ILO Encyclopaedia of Occupational Health and Safety.* Vol. 2, Chapter 42, Heat and cold. International Labour Office, International Labour Organization, Geneva. ISBN 92-2-109203-8 (CD and online) http://www.ilo.org

Wilmore, JH., and DL. Costill. 2004. *Physiology of Sport and Exercise*, Part IV, Chapter 10, Exercise in hot and cold environments: Thermoregulation. Human Kinetics Europe, Leeds, UK.

10 A Good Working Life for Everyone

*Allan Toomingas, Margareta Bratt Carlström,
and Svend Erik Mathiassen*

Photo: Allan Toomingas

CONTENTS

George is a caretaker. After having tried various kinds of work, he has finally found a job he enjoys. He started work 25 years ago in a slaughterhouse, but had to leave his job as a meat-dresser when his arms ached so much that he could not pursue his favourite hobby—darts. Since then he has worked under the pressure of piece work on building sites, where the work always seems to be behind schedule, as a truck driver spending long nights on the road, and sitting the whole days in front of a computer at a call centre selling mobile phone contracts.

It was then that the job as a caretaker came up. Quite the perfect mix of maintaining district heating units, weeding flowerbeds, helping tenants with various repairs and installations, sitting at the computer and ordering goods, and a lot more besides outdoors or indoors according to the season and the weather—a job rich in variation, rarely boring, or doing the same thing twice. Of course, sometimes there may be emergency calls and a lot to do, for example, when there has been a water leak, lifts have got stuck, or parties have become too noisy. Sometimes the job can be heavy, stressful, or uncomfortable, for example, when George has to scrabble around in attics to change the filters on ventilation units or unblock stoppages in the pipes down in the culvert. But those stresses are temporary, and he rarely needs to go to bed worrying about the next day. He is in control of his job and can by and large plan his own working day. The boss has great confidence in George. George has gone down to his standard weight and feels better than he has for many years. He and his colleagues who look after the neighboring residential areas go to the gym in the centre of town twice a week. And last, but not least: he is popular among the tenants, particularly since he started organizing a darts tournament in the club room twice a week. George likes to tell you about what a good life he has—that his job is the best he has ever had. George's job is just right for him—the perfect match.

10.1 WORK IS IMPORTANT

Work is a precondition for the maintenance of society and the great majority of individuals. Society is based on people's work. Chapter 1 described how work has consistently been a dominant part of human existence, and that the location of that work and its role in life have been different during different periods of history.

Apart from providing a salary and support, work also gives the individual a role and a possible sense of coherence in society. It also gives life a structure and a pulse. For the majority of people, the workplace also provides an opportunity for regular social contacts. Work can provide opportunities for using the individual's talents and for professional development. For most people being involuntarily excluded from working life as a result of unemployment or illness is a great hardship.

To feel that you can do a good job and be satisfied with your achievements is characteristic of a good quality of life. To feel needed, but not exploited, is something everyone should experience.

10.2 WHAT IS GOOD WORK—SOME PRINCIPLES

The various chapters of this book have presented proposals for how working life might be designed in order to promote real achievements, good health, and well-being. These proposals follow some basic principles for good work, which are summarized here.

10.2.1 JUST RIGHT, NOT TOO LITTLE, AND NOT TOO MUCH

Rarely is the "untranslatable" Swedish word *lagom** quite as apt as when we are trying to describe what good work looks like from the viewpoint of occupational physiology. In most of the chapters of this book the preconditions for productive and health-promoting work could be summarized using this word *lagom*. Exposures have to be "just right". This applies to a working day that contains just the right tasks, a work pace that is optimal, demands for muscle exertion that are optimal, movements, hot or cold or dry or moist air, psychological demands, and control over one's own work, as well as variation—both physical and mental—that are all optimal. What is optimal may vary between different individuals and at different stages of life. Good work also provides scope for the individual to adapt and develop. It is a challenge in a world of work characterized by specializations, rapid work, slimmed-down organizations, and short lead times to find physical and mental demands that are just right when adjusted to individual needs and capacity.

10.2.2 VARIATION AND RECOVERY

Other key words that have recurred throughout this book are *variation* and *recovery*. People are good at putting up with physical loads and mental stress if these exposures are of short duration and do not recur too frequently. Minor injuries that occur as a result of work will be repaired when exposures cease. But this process of repair

---
* The Swedish word *lagom* comes from an ancient Swedish word meaning "according to the law" and "what we agreed," "suitable," "befitting." Here we translate it as "just right" or "optimal."

presupposes that there is sufficient time and opportunity for recovery (see Chapter 6, Section 6.11). Any imbalance between breaking down (catabolism) and reconstruction (anabolism) leads to chronic problems. Sleep is an important period for repair and reconstruction. A long period of sleep disturbance may therefore be regarded as a serious alarm bell that the body's opportunities for recovery are in the danger zone.

Recuperation and recovery can also occur during waking hours. Breaks can provide an opportunity for recovery. Work can also allow for recovery if the load is varied by, for example, alternating work postures and movements. Structures that previously have been under load can now rest and recover, while other structures are exposed. George's job as a caretaker is an example of a job that can provide good opportunities for recovery because it contains a variety of tasks with different exposure profiles.

Variation in one's work tasks can therefore be seen as a particular attractive opportunity for recovery, since it can be carried out during ongoing work. Unfortunately, we know surprisingly little about what type of variation is the most effective from a physiological and psychological standpoint, how much variation is good in different occupations and for different individuals, and what tasks should optimally be combined, if we wish to achieve an optimal pattern of exertions and recovery.

10.2.3 HEALTH-PROMOTING WORK

It is self-evident to most of us that work should not cause injury or ill health. Great efforts have therefore been made to prevent accidents and work-related disorders of different kinds. The risk of suffering serious injury at work is low in most postindustrial countries. Unfortunately, still many employees are affected by other kinds of disorders that are associated with their work. Examples have been described in the previous chapters.

In addition to preventing *ill health*, there has been an increasing interest in achieving the more ambitious target of promoting and consolidating *good health* at work and through work. The focus so far has primarily been physical training and different kinds of dietary advice. Also, the prevention of addiction, for example, to tobacco and alcohol, is often included in health promotion at work. Avoiding harmful substances is not a way of promoting health, however, but rather a way of preventing ill health. A non-smoker does not improve his health by not taking up smoking. A smoker, on the other hand, curbs his ill health by giving up smoking.

Other important health-promoting factors are access to clean air and pure water, good living conditions, good sleep, natural and cultural experiences, opportunities for personal development, and a good social status. Most of these factors lie outside work and affect what is happening directly at the workplace only to a limited extent, even if there are examples of companies that show concern for their employees' wellbeing and development in a broader perspective. More rarely has the question been asked whether, and in that case, how *the work itself* might be able to promote good health, that is to say, in a narrow sense, boost the individual's capacity and functional ability. Another related aspect is whether work might be able to increase our resistance to injuries and ill health. Those functions and organs where we can most obviously improve health and resilience are those which are exposed at work and which may adapt to different loads through the constant remodeling process described in Chapter 6, Section 6.11, that is, primarily the circulatory and the musculoskeletal systems and mental/psychological functions.

Physical activity is necessary in order to preserve or improve the capacity of the circulatory and the musculoskeletal systems, something that is taken up in Chapters 2 and 6. Physical activity has also been shown to have both a preventive and sometimes therapeutic effect as regards many illnesses, of which a number are both common and serious, for example, high blood pressure, heart disease, diabetes, cancer, and depression (see Chapter 6, Section 6.11).

In the ideal case, the demands of work should be of such a kind and at such a level that they contribute to *improving* the individual's capacity and in this way increasing their work ability. One might imagine that bicycle messengers increase their physical capacity as a result of their work (Chapter 2), or that mentally intensive work for a period leaves the individual better prepared to meet similar demands in the future (Chapter 7). The demands in most occupations in today's working life are, however, such that they do not provide the training effect which should be able to increase capacity or protect against illness. Tasks are seldom designed so that they alternate between periods of high demand for oxygen uptake or muscle force and periods of recovery, which are necessary to build up a good capacity. In order to achieve an adequate training effect, one therefore has to supplement this with physical training outside productive work. Most jobs could, however, be designed so that they contribute to maintaining basic fitness and strength, for example, by avoiding prolonged sitting (see Chapter 6, Section 6.14).

Favourable psychological and social conditions at work also have great significance for both physical and psychological health and productivity. Factors that are particularly important include the opportunity of being able to influence working conditions and achieving stimulation and personal development. It is also important for us to gain support and recognition from people important to us in the social groups within which we move. A good "status" in working life, just as in society, has been shown to be an important factor for good health. The opposite, for example, being bullied or in some other way being made a social outcast, is a serious threat to health. The same applies to being subjected to other people's arbitrary behaviour, threats of impending changes for the worse, or not being able to shake off worries about the future.

The potentials of work in helping promote good health do not apply to all of our biological systems, however. As far as we know, it is not possible, for example, to improve our hearing or the resistance of our auditory system to future hearing impairment from high noise levels. Nor is it possible through work to make a healthy skin more resistant to future eczema or other skin diseases. Nor is it known how it would be possible to improve a healthy person's resistance to chemical poisoning, dust or gases, or to the influence of vibrations or ionizing radiation. The only way is to avoid such harmful exposures.

10.3   MAINTAINING AND INCREASING WORK ABILITY

In Chapter 1, Section 1.4 we described how the concept of work ability has taken the centre stage in working life. Shortcomings in work ability have been noticed among large parts of the population of working age. Shortcomings in work ability are, as described in Chapter 1, Section 1.4, the result of an imbalance between the demands of work and the individual's functional ability, capacity, and working technique. From occupational physiology (and from ergonomics) we can obtain guidelines for how to achieve a better balance.

10.3.1 ADAPT THE DEMANDS OF THE WORK

An important basic principle, if we wish to maintain or increase work ability is to adapt the demands of the job to the capacity and ability of the individual. This does not always need to mean that we lower demands for quantity or quality in the individual's production, even if this may be the answer in situations where there is no other solution. One such example is an understaffed hospital ward, where increasing the number of staff is sometimes the only sensible solution for maintaining the quality of care and the long-term health of staff. Another example is to adapt the work pace, which is a relatively unexploited but powerful measure that has been shown to increase work ability, particularly among older workers.

The demands of work can be reduced, without necessarily lowering production results through suitable technical solutions, equipment, and tools. For example, on a hospital ward one can lower the demands for exerting large muscle forces on the part of the staff by acquiring ergonomically well-designed lifting equipment for patient transfers.

Another important method for adapting the demands of work is to organize it in a way that increases variation in the individual's job. Working methods, equipment, and tools should in themselves allow for variation, but organizational initiatives such as job rotation, job enlargement, and job exchange are presumably more effective sources of variation (see Chapter 6, Section 6.14). The initiatives can be more or less radical and should prioritize those physical and mental functions primarily at risk for overexertion, fatigue, disorders, or injury. The physiological aim of job rotation, for example, is to vary the demands of the job and in this way alternate exposures between several different physical and mental functions, so as, in this way, to favour recovery (see Section 10.2).

If sufficient variation cannot be achieved in the productive work itself, then suitable breaks and pauses should be introduced. In many cases it is possible, particularly in the case of heavy work or work demanding close attention, to reduce exertion without lowering productivity by using short and regular pauses.

Sometimes it is not possible to adapt the demands of work to the individual's capacity by taking reasonable steps, for example, if a truck driver has suffered serious visual impairment. Then the most reasonable solution must then be a change of duties to work where the driver's capacity and ability are adequate. The solution does, however, have its limitations, as some of the commonest causes of reduced capacity, for example, neck and shoulder pain, may imply a reduction in work ability in most occupations where people work a great deal with their hands, including computer work. The same applies to stress-related physical or mental disorders, which may be limiting at most of today's highly effective and goal-oriented workplaces.

There are today unfortunately few ordinary jobs that provide opportunities for an individually adjusted pace of work.

10.3.2 GOOD WORKING TECHNIQUE

Work ability can be maintained and even increased through good working technique. Good working technique optimizes the load, makes for less exertion and less fatigue and reduces the risk of developing disorders and ill health (see Chapter 1, Section 1.4). With good working technique, the individual can utilize opportunities for variation and recovery offered by the task, the work organization and the equipment. Additionally, a good working technique is characterized by people working in a relaxed way, in comfortable work postures, without unnecessary muscle activity or stress. This is facilitated by the worker being fairly familiar with and confident about the task. An individual with reduced capacity as regards certain functions can often regain their work ability by training in a different working technique that makes use of other intact functions. Well-known examples are people with visual impairments who learn to read Braille with their fingers. Persons with "mouse arm" on their right-hand side can train to use their left hand too, and above all learn the short-cut keys on their keyboard so that use of the mouse can be minimized. Older employees can develop a working technique in which the work is organized with more short breaks, which reduces the requirements on fitness and endurance.

10.3.3 INCREASING THE INDIVIDUAL'S CAPACITY

The balance between the demands of work and the functional ability of the individual can also be achieved by increasing the individual's capacity. Many professionals, for example, doctors, actors, and researchers, presuppose a mental and/or social capacity and ability at a level which perhaps only a small section of the population can match. Professions that necessarily presuppose a considerable physical capacity are less common, particularly in industrialized and postindustrial countries. Examples are firefighters, divers, pylon workers, and military and police officers. Here it may be necessary to improve work ability by increasing the individual's capacity in the specific physical functions required. In this way, the short- and long-term physiological effects of the exposure are alleviated. For example, a paramedic or firefighter should be able to carry an unconscious person with only a modest increase in pulse rate and without developing a serious muscle fatigue. This ensures both the safety and health of the paramedic or firefighter and that of the person who is being helped. It is usual that professional groups of this kind are given physical training during paid working hours.

10.4 WHAT THE LAW SAYS

10.4.1 GENERAL ASPECTS

For member states in the European Union (EU), there are a number of directives bearing on working environment and health [EU 1989a, EU 1989b, EU 1989c, EU

1990]. The overarching framework directive describes how the working environment project is to be pursued systematically and in collaboration between employers, safety representatives, and other employees [EU 1989a]. Employers should ensure, for example, that risk assessments are carried out regarding all relevant aspects of the work: physical, mental, and social. If risks are identified, then remedial measures should be taken and followed up. As directives are not particularly detailed, the EU member states have national legislation, often in the form of provisions, which put into concrete form the demands in the directives.

In the United States there are comprehensive working environment regulations in the Occupational Safety and Health (OSH) Act [OSHA 2010a]. They contain no explicit legal demands for ergonomics and physical load at work. In Chapter 5, "Duties," there is a general paragraph that states the employer's responsibility for eliminating risks at work which may lead to an employee dying or being seriously physically injured. There are a large number of mandatory standards in the field of work environment linked to the OSH Act, though none that directly apply to ergonomics and physical load. The federal body that issues these standards, the Occupational Safety and Health Administration (OSHA), has published guidelines on ergonomics for some industries [OSHA 2010b]. These are not mandatory, but refer to the paragraph mentioned above in the fifth chapter of the law. Another important American body, whose activities are regulated in the OSH Act, is the National Institute for Occupational Safety and Health (NIOSH). This is a federal institution, which pursues research, development, and training in the field of the working environment. NIOSH have large-scale programmes that relate, for example, to ergonomics, musculoskeletal disorders, working in cold or heat, and stress at work.

In Asia, Japan, for example, has an "Industrial Safety and Health Act" (Act No 57 of 1972, revised 2006). Article 3(1) applies to the employer's responsibility not merely to prevent accidents, but also in general to create a comfortable industrial working environment in which constantly improved measures are promoted to secure the health and safety of the employees [ILO 2006].

One tool for companies and businesses to live up to the requirements of directive and national laws in practice is to introduce a system for managing the working environment, for example, OHSAS 18001 [BSI 2007]. Other tools are the global (ISO) and European (CEN) standards that often derive from legal requirements. While some CEN standards supporting the EU Machinery Directive (EN 614, EN 1005) give mandatory directions to machinery designers in how to control musculoskeletal health risks, ISO standards are not compulsory, though they do comprise a practical way of fulfilling requirements in laws and national regulations. There are a large number of ISO standards that have a bearing on different types of load at work.

10.4.2 Specific Aspects

The chapters in this book have specified some important EU directives, national Work Environment Acts, and provisions that are relevant to the working conditions dealt with in detail in that specific chapter. These are, generally speaking, focused on protecting the employees *against harm and illness*. This can be read from the formulations of the general aims of the directives and acts, for example:

> "… the employer shall take the measures necessary for the safety and health protection of workers, including prevention of occupational risks …" [EU 1989a, Article 6.1].

> "The purpose of this Act is to prevent ill-health and accidents at work and generally to achieve a good working environment" [SWEA 1978, Chapter 1, Section 1].

The directives and acts do not usually have anything to say about work *promoting health*. In Norway, as an exception, the expectations regarding work are on a higher level. It is stated in the Norwegian Work Environment Act, (Chapter 1, Section 1.1) that: "The purpose of the Act is to secure a working environment that provides a basis for a health promoting and meaningful working situation …. "

There is, on the other hand, some support in the EU directives and the national Work Environment Acts for the need for physical and mental *variation* at work, for example:

> "…adapting the work to the individual, especially as regards… working and production methods, with a view, in particular, to alleviating monotonous work and work at a pre-determined work-rate and to reducing their effect on health" [EU 1989a, Article 6.2d].

> "The employer shall ensure that work which is physically monotonous, repetitive, closely controlled or restricted does not normally occur. If special circumstances require an employee to do such work, the risks of ill-health or accidents resulting from physical loads which are dangerous to health or unnecessarily fatiguing shall be averted by job rotation, job diversification, breaks or other measures which can augment the variation at work." [SWEA 1998, Section 4].

Neither EU directives nor the national Work Environment Acts have anything to say on the subject of *work ability*. There is, however, clear support for the adaptation of the job to the capacities and preconditions of the individual worker, for example:

> "… adapting the work to the individual, especially as regards the design of work places, the choice of work equipment and the choice of working and production methods" [EU 1989a, Article 6.2d].

> "Working conditions shall be adapted to people's differing physical and mental aptitudes." [SWEA 1978, Chapter 2, Section 1].

Regarding *working technique*, the employer has the obligation to ensure that the employees have sufficient knowledge of healthy and safe work:

> "The employer shall take appropriate measures so that workers and/or their representatives … receive … all the necessary information concerning:

> – the safety and health risks and protective and preventive measures and activities in respect of both the undertaking and/or establishment in general and each type of workstation and/or job" [EU 1989a, Article 10.1a].

"The employer shall ensure that the employee has sufficient knowledge regarding:

1. Suitable work postures and working movements.
2. The proper use of technical equipment and aids.
3. The risks entailed by unsuitable work postures, working movements, and unsuitable manual handling.
4. Early indications of the overloading of joints and muscles.

The employer shall further ensure that the employee is given the opportunity of training in a suitable working technique for the task involved. The employer shall also ensure compliance with the instructions given" [SWEA 1998, Section 6].

There are no general demands on the minimal *capacity* of employees in the directives or acts. But the employer has to take this into consideration nevertheless:

"Where he (the employer) entrusts tasks to a worker, he must take into consideration the worker's capabilities as regards health and safety" [EU 1989a, Article 6.3b].

The Swedish Work Environment Authority has promulgated provisions on occupational medical supervision concerning occupations that make heavy demands on the physical capacity of the worker: divers, fire fighters using breathing apparatus, and pylon workers [SWEA 2005]. These professionals need a specific certificate based on a thorough medical examination stating that they are fit for the job.

10.5   TRENDS IN TODAY'S WORKING LIFE

Working life has always undergone change, as mentioned in Chapter 1. New lines of business and new occupations appear and others disappear. Changes are usually slow and occur over many decades. In certain cases they can be dramatic, such as the growth of call centres, which in the past 10 years have become the most rapidly growing sector of the labour market. Another trend in today's working life is the expanding "entertainment sector" which generates new professional groups, from designers of computer and adventure games to wilderness guides and retreat leaders. Other businesses will essentially endure, for example, schools and education, health care, care of the elderly and needy, and other occupations in the service sector, including retail, handicrafts, hotels and restaurants, and office cleaning. The service sector is constantly expanding, and today employs almost 70% of all workers in the EU member states; at the same time the number of workers employed in the production of goods, in agriculture and forestry, and in manufacturing industry is gradually decreasing. Changes are also occurring as regards working methods and equipment. For example, information and communication technology (ICT) is used for an increasing number of tasks at work.

All the changes in the labour market and in working methods naturally have effects on working conditions. Today, >60% of all workers within the EU are white-collar workers with varying levels of skill. Major changes are also occurring in the structure of the working population. People are entering employment later in life in step with the longer period of education among young people. The proportion of older workers is increasing as a result of demographic changes in the population. The pension age is in the process of being raised in many countries. The health of older people has generally improved in each generation, which boosts the opportunities for prolonging working life. Many older people are, fortunately enough, also interested in continuing to work, provided that the job can be adapted to their needs and conditions, for example, as regards working hours. This means that the average age of the working population will increase, and that an increasing number will therefore have age-related reductions in physical capacity and various more or less serious health problems. This presents working life with challenging demands for adjustments.

In many countries there are major groups of inhabitants who were born abroad. They presumably bring with them a great variety of different experiences of working life from their home countries, which should enrich business and working life in their new country. Possible problems for the individual need to be addressed with mutual understanding and adaptation. Linguistic and cultural differences may sometimes make understanding things like safety instructions, for example, more difficult. Current difficulties for individuals of differing ethnic origin in finding suitable jobs, particularly those on a level with their personal competence, should successively diminish as the demand for labour is expected to increase (see below). This should also involve a decline in employment in the gray sector, where the working environment and safety is often deficient and is beyond the supervision of the authorities.

The feared labour shortage, assumed to be a result of demographic changes in the population, should result in everyone being welcome on the labour market—even those people who, for different reasons, have reduced capacity and ability, for example, on the grounds of age, ill health, or limitations in mental or physical abilities. Adapting work requirements to the needs and capacities of different individuals then becomes increasingly necessary to retain work ability among the working population in general.

Working life is becoming increasingly knowledge intensive. "Simple" jobs are disappearing at a rapid rate. Knowledge acquired in compulsory school does not go very far. Requirements for physical capacity and skills are being replaced to an ever-increasing extent by requirements for mental, emotional, and social competence. In each generation, more individuals are expected to gain university or college degrees to fulfill the demands for skills in business. The opportunities of further education and the gradual shift in the demands for skills may, of course, be stimulating for the majority of people. For those who, for various reasons, find it difficult to fulfill the demands, it is important for the actors in working life to find alternative careers.

Another current trend in today's working life, which will presumably continue in the future is the increasing emphasis on individual responsibility. Individuals should take responsibility for being "serviceable" by keeping competent, healthy, and strong. They should exercise and eat sensibly. They should update their CVs and their qualifications. They should cultivate their networks and acquire a personal trainer and coach. On the one hand, this trend provides the individual with opportunities for stimulation and development, and greater opportunities for control and management of one's own work. But if the stress on the individual is combined with the trend towards demands for greater confidence and efficiency, as described above, then this development can become a stressor and a source of insecurity for the individual who is not stimulated by a drive to "invest in yourself" or is lacking in self-confidence. An individual who does not succeed easily gets left behind. Warning signals can be seen, for example, among many school students with stress and pains who view a future working life with concern. It is, therefore, an important challenge for the community to design new welfare and social security systems that suit the new conditions on the labour market.

Within the manufacturing industry one can see a development in the Western world towards most large companies outsourcing parts of their production or their support functions to specialized subcontractors—preferably in other countries—and rationalizing the production that remains towards greater elements of automation and computerization. Among both subcontractors and parent companies, the choice of different tasks in this way becomes more limited. Those tasks that are on offer are very similar as regards physical demands and therefore provide only limited opportunities for varying work postures, movements, and force development. A further trend in both industry and the service sector is to introduce neo-Tayloristic production systems with a strong focus on avoiding non-value-added time, which is considered waste. This probably leads to less scope for variation and recovery in that spontaneous stoppages at work have disappeared, as well as to more short cycle, repeated operations.

More and more work is done with the help of ICT. Approximately 70% of the working population in many countries today use the computer in their work, and the proportion that have a computer as their main equipment is increasing (see Chapter 6). At a rough estimate, 25–35% of all hours worked in many countries are now spent on a computer, and the proportion will presumably increase.

The computer has thus become our commonest tool. It is a valuable aid that facilitates and streamlines work and which has meant a radical development in most enterprises in society. The opportunities for communication and contact have also been developed considerably, for example, through mobile phones, text messaging, and e-mail. One example of modern communication technology in combination with computers is the huge growth in customer service centres (call centres) which started to expand in the 1990s. Increasingly advanced and complex services that require higher education will be provided via this form of customer communication. Examples are information on health and pharmaceutical preparations, legal and financial advice, and various forms of technical support. Even public authorities are increasingly transferring their customer contacts to call centres, for example, the tax authorities, the social insurance office, and police service. This expansion of call centre activity, like the expansion in many other lines of business, will bring a further increase in the number of workers with prolonged, low–intensity, and constrained sedentary work. Physical inactivity may in the future become one of the greatest problems in the working environment [European Agency 2005].

The development within ICT also results in work becoming "unbounded," or flexible, in time and space when it is no longer tied to a particular workplace, as one can work on the computer even at home or when traveling. This boundlessness may increase the individual's freedom to decide when and where they do their work, which may make it easy for work to be combined with other tasks in life, for example, looking after ailing parents. Other people can continue to work from home when their workplace has moved to another town or when they themselves have chosen to move away. It is, however, a question of avoiding working hours becoming infinitely long when, for example, one takes work back to the home computer. Also, this development means that mental and physical exposures at work and leisure become more similar; life as a whole offers less variation.

Even the boundaries within workplaces have been eradicated, as open office landscapes have become more common in many sectors of the labour market. Sometimes workers do not even have their own desks, but arrive at work and sit in anywhere that which is temporarily vacant. This makes demands on adaptability both as regards equipment and the individual.

Another trend is the move towards a 24-h society. People expect services to be provided round the clock, every day of the year. This form of boundlessness in time has consequences for staff working hours. A similar boundlessness also affects other large groups in working life who are expected to be online and available via e-mail and mobile phone in their leisure time, including holidays. The opportunities for mental recovery then have to be emphasized, so that availability is not so comprehensive and intrudes on private life and leisure time.

10.6 WORK–LIFE BALANCE

A good job provides an optimal combination of stimulations, challenges, development, and safety that is just right, and is characterized by a balance between the demands of work and the individual's working technique and capacity. In a working life that one-sidedly stresses the responsibility of the individual, there will only be room for those individuals who have great self-confidence, good capacity, and a well-developed working technique. This may provide problems at times when the capacity is utilized to the full and may even be found wanting, for example, during parenting, just as during periods of ill health or when one grows older.

Good work should stimulate everyone according to their preconditions towards a positive development of their skills and work ability. Good working life should also be "permissive" in its demands, so that even those people with limited capacity and less than optimum working technique are able to do a valuable job with good work ability right up to retirement age, while maintaining good health and quality of life for the duration of their life.

That would be a good working life for everyone!

REFERENCES

BSI. 2007. *OHSAS 18001:2007.* British Standards Institution. www.bisgroup.com

EU. 1989a. Directive 89/391/EEC—*On the Introduction of Measures to Encourage Improvements in the Safety and Health of Workers at Work.* http://eur-lex.europa.eu/

EU. 1989b. Directive 89/654/EEC—*Concerning the Minimum Safety and Health Requirements for the Workplace.* http://eur-lex.europa.eu/

EU. 1989c. Directive 89/655/EEC—*Concerning the Minimum Safety and Health Requirements for the Use of Work Equipment by Workers at Work.* http://eur-lex.europa.eu/

EU. 1990. Directive 90/270/EEC—*On the Minimum Safety and Health Requirements for Work with Display Screen Equipment.* http://eur-lex.europa.eu/

European Agency. 2005. *Expert Forecast on Emerging Physical Risks Related to Occupational Safety and Health*. Luxembourg: European Agency for Safety and Health at Work.

ILO. 2006. *Industrial Safety and Health Act* (Act No. 57 of 1972). Unofficial translation from Japanese to English by *The International Labour Organization*. http://www.ilo.org

OSHA. 2010a. *OSH Act of 1970*. Occupational Safety & Health Administration. www.osha.gov/pls/oshaweb/owadisp.show_document?p_table = OSHACT&p_id = 2743

OSHA. 2010b. *Regulation (Standards—29 CFR)* http://www.osha.gov/index.html

SWEA. 1978. *Swedish Work Environment Act* (last revised in May 2011). The Swedish Work Environment Authority. http://www.av.se/inenglish/

SWEA. 1998. *The Swedish Work Environment Authority's Provisions on Ergonomics for the Prevention of Musculoskeletal Disorders*. AFS 1998:1. http://www.av.se/inenglish/

SWEA. 2005. *The Swedish Work Environment Authority's Provisions on Occupational Medical Supervision* AFS 2005:6. http://www.av.se/inenglish/

Index